Thought Experiments

Nenad Miscevic

Thought Experiments

Springer

Nenad Miscevic
University of Maribor
Maribor, Slovenia

ISBN 978-3-030-81081-8 ISBN 978-3-030-81082-5 (eBook)
https://doi.org/10.1007/978-3-030-81082-5

This Springer imprint is published by the registered company Springer Nature Switzerland AG
The registered company address is: Gewerbestrasse 11, 6330 Cham, Switzerland

Preface

This book is an introduction to thought experiments and theories concerning them.[1] What is a thought experiment? Start with an example.

Imagine a scientist, call her Mary, who lives in a black and white room and has never seen real colors. She specializes in the neurophysiology of vision and acquires *all the physical information there is to obtain* about what goes on when we see various things. One day she is released from her black and white room and for the first time sees blue, red, and other colored things. Will she learn something new? The obvious reply is Yes. But then, physical knowledge is not all there is to the knowledge about the world; perhaps physical facts are not the only facts there are in the world.

The story and its long-reaching morals, all due to Frank Jackson (1982), present one of the most famous thought experiments in the philosophy of twentieth century. The author invites us to build a scenario in our imagination and react to it. Assuming you went through the story, you have probably ended up with an opinion or at least with being strongly attracted to the idea that Mary learns something new. This immediate result of performing a thought experiment is usually called "intuition". Thought experiments, in general, are such imaginative exercises of thought, ubiquitous in philosophy and also appearing at crucial joints in science, most famously in the work of Galileo and Einstein (for more examples, see the next chapter, "Introduction").

Thought experiments and intuitions are now being intensely and extensively discussed in theory and philosophy of science, as well as in various other branches of philosophy, theoretical and practical. There are popular introductions and presentations of topics from physics and philosophy through thought experiments and lots of papers and books about them.[2] The methodology is being applied to the burning

---

[1] The biggest thanks go to Jim Brown, who has introduced me to thought experiments decades ago and has supported me in many ways.

[2] In philosophy, the typical examples are Cohen, M. (2004), and Schick, T. and Vaughn, L. (2012); for science, one has illustrated books, with titles like*Einstein for Beginners* or …*for Dummies*, e.g. Schwartz and McGuiness 1979., or Calle, 2005.

For general information on thought experiments, see the excellent *Routledge Companion*, (Stuart et al. (2018).

issues of the moment; see for instance Amgad M., (preprint, 2019) on a highly virulent emerging viral epidemic, imagined along the lines of actual SARS-CoV, H5N1 Influenza, and Ebola pandemics.

We would like to offer an introduction to the topic, combining the overview of the field, and some proposals of our own about understanding thought experiments. Here is a short preview of the book. In the introductory chapter, we go through a handful of examples from science and philosophy and then turn to definitions; among several proposals, we choose one and add some comments to the initial proposal. Then we place thought experiments on a wider map, noting their similarity with imaginative exercise in fiction, and other, less expected areas; most of this work has not been done before in a systematic way.

Chapter 2 turns to the next two topics: first, the role of thought experiments, and second, their typical internal structure. We defend the view that thought experiments are extremely useful in science and indispensable in philosophy. But in continental philosophy, at least in the most typical authors, you find no thought experiments, the reader might point out. Well, they are replaced by their nearest kin, the fictional stories, or artistic representations, that are treated practically in the way traditional philosophy and its analytic continuators treat thought experiments. Continental philosophers do experiment in thought, just as their colleagues have been doing, but only use a different name for the activity.

The next issue is the internal structure of thought experiments. Philosophers have noted that they do have a structure, beginning with the story or scenario, accompanied with a question, continuing with building an internal model, and producing an intuition. Authors like Chisholm (1987) and Snowden (1992) have pointed out that scientists and philosophers typically vary their scenarios and try to elicit a general conclusion from a series of such varied stories.

Rawls has then concentrated on the efforts of philosophers to bring their thought-experimental conclusions in balance with other accepted views, commonsensical or scientific or both.[3] This final phase he called Reflective equilibrium.

Here, we first systematize the stages, which are often sketched by philosophers in different and unsystematic ways; then we use the final systematization thus obtained in the last two chapters to do some new work in theorizing about TEs, in particular, their weaknesses and their typical philosophical history.

Chapter 3 is an overview of the extant theories of thought experiments and of intuitions. Unfortunately, we have to be brief, so we present just the main ideas dominating the playground. We start with empiricist-inferentialist proposals, historically very prominent both in relation to scientific and more philosophical thought experiments. Its proponents, from Mach to Norton, see thought experiments as arguments, enriched with illustrative, not very relevant details. We then pass to the opposite extreme, the radical Platonist alternative. Next, we turn to more moderate proposals, first a conceptualist, then a Kantian proposal, and thirdly a recent proposal due to

[3] Originally in 1971; see now the (1999) edition.

Tim Williamson, stressing the role of counterfactuals and counterfactual imagining. The chapter concludes with a brief pointing to the last, fourth option which we prefer, the mental modelling theory.

Chapter 4 develops the mental modelling theory alternative and adds a perspective on intuitions as a "voice-of-competence" phenomenon. We introduce our own proposal concerning the duality of model building, i.e. imagining from a neutral, third person-perspective and simulating from the first-person perspective. The mental simulation has been studied, but not very much in the context of thought-experimenting; we offer a unified but differentiated picture that corresponds to important differences within the set of thought experiments. We argue that the mental modelling theory and the "voice-of-competence" proposal can account for most, perhaps even all, puzzling phenomena tied to thought experiments and intuitions.

Throughout the book, we appeal to the internal structure of thought experiments and stages that characterize and define it; an appeal that is not being made in the literature. In the last two chapters, we apply our interest in stages and our mental modelling proposal to two topics that do deserve special treatment. Chapter 5 raises an issue that is not, in this form, being raised in the literature, namely how to understand the history of thought experiments from the very long one in the case of Descartes' Evil Demon to shorter but also very rich ones with Einstein's or Bohr's experiments.[4] We propose that in a typical such history, various stages become prominent in various confrontations and that one can systematize the history by tying it to the stage structure. A lot more needs to be said and done, but this is, to our knowledge, the first attempt to systematize the issue in the literature; the rest, namely the actual analysis of the very rich high-quality material, from Descartes to Einstein and Gettier, is best left for the future.

Finally, Chapter 6 addresses a recent movement critical of thought experiments, the so-called "experimental philosophy" that has been at the center of discussion of thought experiments in the last two decades. The leading idea has been to test naïve intuitions from thought experiments and apply the usual psychological methods above all statistical techniques in doing it. I argue that the attention to stages might show a way out for the proponent of thought-experimenting: the weaknesses that experimental philosophers detect in thought experiments are heterogeneous and belong to different stages, so they can be addressed in a differentiated fashion. This would lead to a kind of "self-help" program for thought-experimenters and might help them to a significant extent. The chapter concludes with a brief discussion of a question not addressed by experimental philosophers, but raised a decade and a half ago in the literature: are scientific thought experiments better off than philosophical ones? It concludes in favor of equality between the two kinds.

---

[4] The only exception I am aware of is the recent excellent chapter by Alisa Bokulich and Mélanie Frappier in the (Stuart et al. eds., 2018) volume, which focuses upon one important sub-issue of the history issue, namely the identity of thought experiments through time.

This book owes its existence to a great number of people. Let me start with James Robert Brown, continue to Michael Stuart, Tim Williamson, and Friderik Klampfer, and conclude with Marko Grba, a physicist and philosophy student who has helped me with the chapter on Einstein.

Maribor, Slovenia Nenad Miscevic

Contents

Chapter 1
What Are Thought Experiments?

1.1 Introduction

1.1.1 Two Famous Thought Experiments

Suppose you have a friend who is very naïve when it comes to physics. He follows his commonsense view, uninformed by high-school physics, that heavier bodies fall faster than the light ones: the heavier the body is, the faster it falls. That's it. You can make him start thinking with a very simple question. "Imagine you tie together a heavy body, say a cannon ball, and a very light one, say a musket ball. If you are right, we have a problem. The new, composed body is faster than the heavy one, the cannon ball, alone, so it should fall faster. But on the other hand, here is this slow body, the musket ball, tied to it. It is small, therefore slow, so it will slow down the composite thing. So, the composite body will fall more slowly than the cannon ball, and more quickly. Isn't this a bad problem for your view?" If you had made a bet with your friend, the question might bring you a pack of beer cans.

In fact, the question is famous, and you probably know it. It is the kernel of Galileo Galilei's attack on the Aristotelian commonsense. His interlocutor, Simplicio proposes the idea, and Galileo himself, represented by the character called Salviati bids Simplicio to consider what happens when we "unite" the heavier and thc lighter body together: "If then we take two bodies whose natural speeds are different, it is clear that on uniting the two, the more rapid one will be partly retarded by the slower, and the slower will be somewhat hastened by the swifter" (1914, p. 158). It seems then that the combined system will be slower than the heavy body a1one. But, on the other hand, the combined system is heavier than the heavy body alone, so it should fall more quickly than the heavy body alone. The supposition that weight is relevant to speed leads to absurdity. So, Salviati can conclude that the difference in weight is irrelevant to the speed. Let us call this the intermediate conclusion. From this conclusion, Salviati jumps to the final conclusion: all bodies fall at (approximately) the same speed.

N. Miscevic, *Thought Experiments*,
https://doi.org/10.1007/978-3-030-81082-5_1

This story is usually taken to present one of the most famous "thought experiments" in the history of human thought, dating from roughly the year 1630. The experiment is not real, but imagined. Galileo apparently did some real experimenting from the Pisa tower, but the thought experiment stayed more famous than the real ones. Although the experiment is famous indeed, it gives no positive suggestion as to the calculation of the speed. We need a more positive example to work with.

Here is a beautiful imaginative thought experiment due to Dutch mathematician Simon Stevin, published in 1586 in his *Book of the elements of the art of weighing*, through which he derived the condition for the balance of forces on inclined planes. (See a fine account in Brown, 1991, p. 4.) It starts with a general issue: we are invited to imagine a prism-like pair of inclined (frictionless) planes with linked weights such as a chain draped over it. How will the chain move? Maybe, it will remain still, or it will move in the direction of greater weight or of the steeper slope. Stevin now proposes that we imagine we link the chain at the bottom. Here is then the resulting situation.

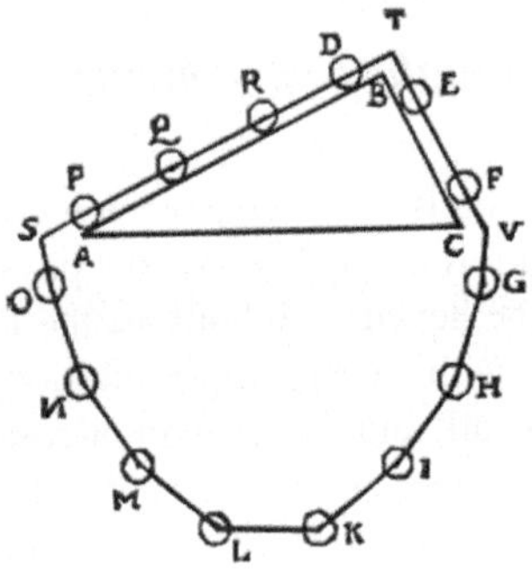

Now, if the force on the left were not balanced by the force of the right, the loop would rotate. But the imbalance would, of course remain, no matter how long it rotates. Thus, we would have made a perpetual motion machine. But this is impossible, Stevin rightly assumes. So, no motion, neither to the left, nor to the right is possible in this case. So, the chain will remain still. Conclusion: the chain is already balanced in the initial situation, and it would not move then as well.

Let me shorten "thought-experiments" to "TEs". The history of physics is rich with important TEs; the ones equally famous as the Galileo's pioneering one, are Einstein's inventions that announce the birth of Theory of Relativity, special and general one. The first, not very much worked out, dates from the time when he was only sixteen.[1] He tried to imagine what it would be like to chase a beam of light, and finally catch it. What would the chaser see? If we imagine the catching like surfing, what is seen should be a stationary wave; but with light this is impossible, since it exists only in moving. There is no stationary light. This mini thought-experiment

[1] For a useful short exposition see Peacock, K.A., "Happiest Thoughts Great Thought Experiments of Modern Physics" in Stuart, M, Fehige, Y. and Brown J.R. eds. (2018), 211–241. For an advanced interpretation and discussion see Norton (2013).

gave Einstein the idea that speed of light is something very special, and that physical theories of his time are impotent when it comes to understanding it. We shall recount his later thought experiments, from his famous papers from 1905 later. We shall also be mentioning some other famous cases from the history of physics.

Our examples nicely illustrate the contrast between destructive and constructive thought experiments (Brown 1991:33). Galileo's one is destructive, ending with a negative morals, namely that the Aristotelian assumption about free fall is false. (It does also support a positive conclusion, that all bodies fall with the same speed, but its main point is criticism of Aristotelians). Stevin's is constructive, proposing a definitive answer to the initial question.

Let us now widen the domain a bit, passing from physics TEs to mathematical ones. As J. R. Brown has convincingly argued in his *The laboratory of the mind*, mathematical insights in many cases parallel TEs. Stevin's TE is akin to problem solving in geometrical by visual imaginative means. Let us start with the examples featuring simple geometrical configurations, like in the famous geometrical tasks in Plato's *Meno.* The simplest case might be claims like the one that by folding a square over one of its diagonals, the opposite sides will come to coincide, In such geometrically «thought experiments» the cognizer is invited to imagine some spatially salient situation and is lead to "see" the necessary regularities which govern it:

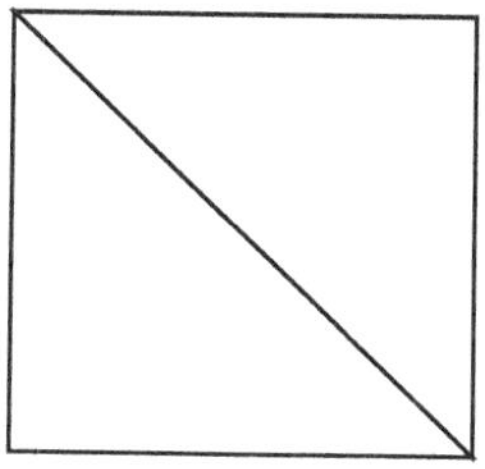

The cognizer might arrive at the insight that the diagonal cuts the square in two equal parts. From such simple insights ('intuitions'), one might pass to more sophisticated ones, all the way to true discoveries due to geniuses like Hilbert. Here is Hilbert illustrating the paradoxical nature of infinity through a nice imaginative example:

> We now assume that the hotel has infinitely many rooms numbered 1, 2, 3, 4, 5, ... and that each of the rooms is occupied by a single guest. All that the manager has to do in order to accommodate a new guest is to make sure that each of the old guests moves to a new room with the number one unit larger. In this way room 1 becomes available for the new guest. One can of course make room for any finite number of new guests in the same manner; and thus, in a world with an infinite number of houses and occupants there will be no homeless. (Hilbert 2013, p. 730).

What kinds of activies count as mathematical TEs? We should look for activities that bring new insights, but not through formal, logical proof procedure. The literature is therefore favoring visual, imaginative thinking. For instance, a recent authoritative overview "Thought Experiments In Mathematics", in *Routledge Handbook of*

Thought Experiments, written by Irina Starikova and Marcus Giaquinto mentions purely abstract calculation, but then narrows the area down to a kind of thinking which involves active use of visual imagination and goes beyond the application of mathematically prescribed rules, as a way of answering questions or overcoming obstacles (2017, p. 275). Implicitly the same focus is taken by other authors, like J. R. Brown in his *Philosophy of Mathematics, A Contemporary Introduction to the World of Proofs and Pictures*,

1.2 More Examples—From Science to Philosophy

Let me offer more examples from science and philosophy. The history of physics is rich with important thought-experiments; the ones equally famous as the Galileo's pioneering one, are Einstein's inventions that announce the birth of Theory of Relativity, special and general one. The first, not very worked out one, dates from the time when he was only sixteen. He tried to imagine what it would be like to chase a beam of light, and finally catch it. What would the chaser see? If we imagine the catching like surfing, what is seen should be a stationary wave; but with light this is impossible, since it exists only in moving. There is no stationary light. This mini thought-experiment gave Einstein the idea that speed of light is something very special, and that physical theories of his time are impotent when it comes to understanding it. We shall recount his later thought experiments, from his famous papers from 1905 later. We shall also be mentioning some other famous cases from the history of physics.

Let me now pass to the area in between science and philosophy, and give you a famous example born in the dialogue between philosophy of mind and computationalist cognitive science, Searle's so-called Chinese room TE in (philosophy of) cognitive science, from 1980 (Searle 1980). It figures a person, closed in a room, who thinks of himself as ignorant of Chinese, but is endowed with a necessary equipment, dictionaries and symbol manipulating rules, for translating from Chinese into English and back. "Imagine that people outside the room send in other Chinese symbols which, unknown to the person in the room, are questions in Chinese (the input). And imagine that by following the instructions in the program the man in the room is able to pass out Chinese symbols which are correct answers to the questions (the output). The program enables the person in the room to pass the Turing Test for understanding Chinese but he does not understand a word of Chinese" (Searle 1999, pp. 115–116). The allusion to the Turing test concerns the idea, very popular in cognitive science, that a device understands and thinks if its "linguistic behavior" is indistinguishable from human behavior, it correctly asks and answers the questions and so on.

Let me add one more example, a constructive TE from the study of language (linguistics and philosophy of language). Many ordinary words, like "water", or "cat" refer to natural kinds. But what exactly does determine their reference and how? In case of water, it could be a folk description (transparent tasteless liquid, common in our oceans and lakes, etc.), or the information about its composition, like "H_2O". Which one is it? Hilary Putnam has famously introduced his Twin Earth story, a "science-fiction example" as he calls his TE (or a series of related TEs). The reader is asked to suppose that somewhere in the galaxy there is a planet we to be called 'Twin Earth', very much like Earth; in fact, people on Twin Earth even speak English. In fact, apart from the differences we shall specify in our science-fiction examples, the reader may suppose that Twin Earth is exactly like Earth. However, the oceans and lakes and seas of Twin Earth contain XYZ and not water, that it rains XYZ on Twin Earth and not water, etc. 223.

Twin Earthlings speak (approximately) English and use the word "water" to refer to the liquid in their oceans and lakes. Now, when a Twin Earthling, ignorant of chemistry, says "This is a glass of water", to what does the word "water" refer to in his parlance, to the liquid which is in fact H_2O or the liquid which is in fact XYZ? The question is meant to test our referential intuitions. And indeed, the subjects very often answer that it is XYZ. So, it is not external properties of the liquid, but its inner "nature", even if the speaker is ignorant of its essence that determines the reference in ordinary parlance. On Earth, in contrast, the word refers to H_2O, and does and did so even in the mouths of speakers ignorant of chemistry, or speakers talking before chemists had figured out the compositions of water.

Next, let me mention a purely philosophical theoretical TE, due to Edmund Gettier (1963), testing the popular theory that to have knowledge amounts to having a belief that is true and justified. In one of its scenarios, Smith is competing for a job. He was told by the president of the company that Jones will get a job. Now comes a bit artificial part of the scenario that usually confuses freshmen student. Smith counts money from Jones's pocket, finds ten coins, and reasons from this fact: "The man with ten coins will get the job". In fact, the future is much rosier for Smith than he thinks. He is the person who will get the job, and, lo and behold, he also has ten coins in his pocket. So, his belief "The man with ten coins will get the job" is true. It is also justified, since Smith heard the news about Jones, who also happens to have ten coins in his pocket, from otherwise reliable president. But, does Smith know that the man with ten coins will get the job? Many subjects come with a negative verdict. Something is missing, so the popular theory that truth and justification suffice for knowledge cannot be correct.

Of course, TEs are not limited to theoretical matters. Some famous ones have been about practical philosophy. Plato's *The Republic* features an early and obvious example, the Ring of Gyges. In order to show that those who practice justice do so unwillingly and from want of power to commit injustice, a participant in conversation, Glaucon, proposes that we imagine a person who has, due to a ring that makes him invisible, acquired full power to commit injustice, and observe in imagination what such a person will do. He recounts the story of Gyges the invisible man, who according to the story did indeed commit a lot of unjust actions, including theft, rape and murder. If you had the powers bestowed by invisibility, would you do the same, he ask the interlocutors. Would you be able to resist the temptations? He expects a negative answer, which he then generalizes: like you, my dear interlocutor, people in general would behave badly if not constrained by fear from punishment.

However, *The Republic* has much more to offer. Towards the end of Book Two, Socrates invites his interlocutors to observe" in logos" the origin of a polis, in order "to see also the origin of justice and injustice in it [369a]." (Later, the same activity will be described as observing "in story" how the polis is developing). The actual construction of the "story" is guided by Socrates' appeals to imagination. He invites his interlocutor to imagine an arrangement, for instance the community of goods. The interlocutor does as told, and then is asked whether he thinks the arrangement is just (or useful for the parties, or even merely acceptable). The usual answer is the famous (or notorious) "Oh yes, Socrates!", "Indeed, Socrates". Then they go to the next arrangement to be imagined, e.g. community of children. "What would happen; would it be just?" "Indeed it would"; and so on. We end up with a mega-TE encompassing at least two hundred pages, and resulting in a model for an ideal state.

Finally, let us remind ourselves of a famous proposal of a political TE due to John Rawls; to put it in his own words it is a "thought experiment for the purpose of public- and self-clarification" (Rawls 2001, p. 17). The situation is the one in which we choose general principles of a just society. Rawls notes that people, if they know their status and capacities, choose egoistically giving priority to qualities they in fact have:

> For example, if a man knew that he was wealthy, he might find it rational to advance the principle that various taxes for welfare measures be counted unjust; if he knew that he was poor, he would most likely propose the contrary principle. To represent the desired restrictions one imagines a situation in which everyone is deprived of this sort of information. (Rawls 1999, p. 17)

However, in the situation imagined one has information about normal general social patterns. All sorts of kinds and properties will turn out to be relevant. If I, a male, did not know whether in the final situation I shall be male or female, it will be rational for me to choose equality of genders, just in case I end up a as female. The ignorance is metaphorically described as a veil, the veil-of-ignorance). The whole scenario is famously called "The Original position". This "procedure of construction" as Samuel Freeman usefully describes it, is part and parcel of the social contract tradition (Freeman 2007, Introduction). The veil of ignorance is "a guide to intuitions" (1999, 139), and this instrument gets honed in the course of development of the discussion in the main works in which it appears.

Let me conclude this section by introducing another contrast, this one having to do with the size of the TE, as presented in the text. All our examples until now have been of standard, "normal" size, half a page to a page in written text. But there is an important tradition of much bigger TEs. For example, in political philosophy, when Plato wants us to imagine a just state, the imagining turns to one issue after another: would such-and-such division of power be just, would such-and-such constellation of military power be acceptable, would such-and-such arrangement for reproduction and child care be all right? A series of normal-size TEs, each on one topic, is then integrated into a book size whole, *The Republic*, in our example, a macro-TE, as we might call it. So, there are normal size TEs and macro-TEs. What about very short proposals to imagine something? Ludwig Wittgenstein's opus is full of such micro-proposals; and here is a famous one. We are invited to imagine a language that "is meant to serve for communication between a builder A and an assistant B. A is building with building-stones: there are blocks, pillars, slabs and beams. B has to pass the stones, and that in the order in which A needs them. For this purpose they use a language consisting of the words "block", "pillar", "slab", "beam". A calls them out;—B brings the stone which he has learnt to bring at such-and-such a call.—— Conceive this as a complete primitive language. §2, *Philosophical Investigations* (see References). One of the important points of the TE is to stress the role of linguistic practice and of acts like commanding in the functioning of language, in contrast to the views that have been stressing the representational function of language plus the allegedly central role of naming and reference, and analyzing the abstract properties of propositions presumably expressed by the sentences of language. I propose that we treat such proposals as very short TEs, call them "micro-TEs".

1.3 What is a TE?

We now need a characterization of TEs, and there is a plethora of proposals in the literature. Let us begin with Einstein, the great thought-experimenter. In his book, jointly written with Leopold Infeld, he is keen to keep his imaginary exercises as close to real experiments as possible. He calls experiments "created by thought", like the ones we were describing, "idealised experiments". According to him and Infeld such experiment present the following further characteristics:

- they are "fantastic," they can neither be derived from experiment nor actually performed;
- they are consequently created "by thought" and even by "speculative thinking";
- they are however "consistent with observation" and lead to "a profound understanding of Einstein and Infeld (1938:214).

So, the definition, or least, a good characterization of TEs should make clear their connection with laboratory experiments. Einstein and Infeld point to idealization as a bridge between the two: TEs are like idealized lab experiments. This is a good point to keep in mind.

Consider next more recent proposals by historians and philosophers of science. In her historical introduction to studies of ancient candidates for TEs Sophie Roux suggests the following:

> ..., if we try to determine what thought experiments are, we will most probably end up with the following three characteristics:
>
> they are counterfactual, they involve a concrete scenario, and they have a well-determined cognitive intention. (Roux 2011: p. 19)

We shall keep in mind the counterfactual character of typical thought-experimental proposals. We quoted above Einstein's formulation of his famous TE: "If I pursue a beam of light with velocity c ..."; of course, I cannot do it, nor can anyone else.

The scenario is strongly counterfactual. To mention another TE, Searle's so-called Chinese room TE in (philosophy of) cognitive science, from 1980, is less impossible in reality.

Other thought-experimental proposals, Galileo's and Stevin's, invite us to imagine a scenario that might be a possible one, without claiming that it is in any sense real. It such and such were the case, something else would have been the case as well. This is exactly what counterfactuals supply for us according to the standard understanding.

We shall abide by the idea of the scenario being concrete. Let us note that all scenarios proposed in the famous examples are such; more about it will come in the following chapters. And we shall have things to say about cognitive intention; obviously, it should be as close to intention of real experiments as possible. Sören Häggqvist thus characterizes TEs as "hypothetical cases intended to function as experiments, in the following sense: they aspire to test hypotheses or theories (Häggqvist 2009, p. 57). Consider now two more proposals. In his pioneering work *The laboratory of the mind* James Robert Brown suggests that "/T/hought experiments are performed in the laboratory of the mind." and goes on to note that "/B/eyond that bit of metaphor it's hard to say just what they are." (Brown 1991, p. 1). His next suggests that "we recognize them when we see them: they are visualizable; they involve mental manipulations; they are not the mere consequence of a theory-based calculation; they are often (but not always) impossible to implement as real experiments either because we lack the relevant technology or because they are simply impossible in principle, as when frictionless planes or a universe devoid of all matter are involved. Brown, *Laboratory of the mind*, Brown (1991, p. 1). The definition of TE in *Stanford Encyclopedia*, due to him and Y. Fehige, introduce more elements:

> Thought experiments are devices of the imagination used to investigate the nature of things. Thought experimenting often takes place when the method of variation is employed in entertaining imaginative suppositions. *Stanford Encyclopedia*

The most elaborate proposal I was able to find is due to Tamar Szabo Gendler

a. Thought-experimental reasoning involves reasoning about a particular set of circumstances (which may be specified in more or less detail), described at a greater level of specificity than that of the conclusion.
b. The reasoner's mode of access to the scenario is via imagination rather than via observation.
c. Contemplation of the scenario takes place with a specific purpose: the confirmation or disconfirmation of some hypothesis or theory (Gendler 2004:1155).

Let me now propose my own variant of characterizing TEs, quite close to proposals made by Roux and Szabo Gendler, but stressing more the actual process of experimenting. I agree with the importance of counterfactual scenarios and of imagination, and with the testing of theories or hypotheses as the goal. However, I want to bring the characterization of TEs as close to the standard characterizations of lab experiment. So, let me note that scientists, when talking about lab experiments, stress the role of design, which determines which variables in the given situation are to be taken into account. They note that normally experiments follow after the decisions about design are taken, and, if all goes well, they end up with a result, a "verdict" that can be then interpreted as supporting the initial hypothesis or theory.

Similarly, a TE is a process that starts with a design, which involves the determination of the goal(s), in particular the thesis/theory to be tested, and the construction of a scenario to be considered.

It then proceeds with the presentation of the scenario thus constructed to the experimental subject (either the author of the scenario, or an interlocutor), whose mind will be, in Brown's apt metaphor, the laboratory in which the testing will be done.

On the side of the experimental subject, the experiment then continues with the typically imaginative contemplation of the scenario plus some piece of reasoning, culminating in the decision ("intuition") concerning the thesis/theory to be tested.

Once this result is achieved, it can be, and often is compared with results of other experiments in the vicinity.

Take the simple example, Stevin's inclined plane TE. The scenario is, of course, first constructed and then presented to the audience, or to oneself, if we have a lonely creative experimenter. The presentation ends with a specific question, "How will the chain move?" Once the question is understood listener begins to imagine the scenario, and asks herself the question relating it to the scenario imagined: indeed, will it move to the left or to the right? Complicated things might be happening below the conscious level, but normally the listener comes to some immediate spontaneous answer. Suppose, she suddenly realizes that if the chain moves, it will not stop moving, since the configuration above the prism stays the same; if this configuration provokes motion, it will not stop doing so (I am following Brown's beautiful analysis in *The Laboratory of the mind* (Brown 1999, p. 4), to show that my description fits the historical cases as normally understood). She is attracted to this option and quickly assents to it; this seeming, attraction and assent are normally described as "intuition". But then comes a piece of reasoning: "I know that perpetual motion is impossible. So, the chain would not even start moving; if it did this would result in an impossibility. The verdict is then accomplished: the chain would stay at rest (Alternatively, like in Brown's story, the subject might start by considering all three options, eliminate the left and right motion, and stay with the rest hypothesis.) Generally, a successful TE usually culminates in intuition or in some immediate consequence of it, like in our example.

Scientists, philosophers and teachers know that this is not the end of the story: one can and should vary the story and generalize the result, and then test the intuition and generalization, comparing them to other spontaneous intuitions and generalizations, or even to information from psychology of belief-formation. When teaching Stevin to high-school kids one might be tempted to introduce variations, to check their understanding. What if we replace chain-and-balls with a thick, very slippery plastic tube? Would it start moving? The pupil should understand that the morals of Stevin's story is rather general, valid not only for chains. Finally, Stevin and his successors have integrated the result of the TE into the general framework of physics of their time. Einstein has done the same with his TEs. The result should be brought into some kind of equilibrium with the rest, and for some TEs this reflective equilibrating might be a considerable task. (Think of Galileo's negative result; it just would not fit Aristotelian physics!). We shall in the sequel be stressing the importance of taking into account the whole 'biography' of intuition, from the performing of the TE all the way to the final, equilibrated result coming from comparisons and corrections. But in the first three chapters, we concentrate upon the first stages, the ones constituting the performing of the TE.

1.4 TEs and their Relatives

Until now we have been considering TEs in strict and literal sense. So much about the internal build-up and "biography" of a typical TE. We now turn to its "surroundings", to its "closest relatives", so to speak. There is, namely, a wide range of intellectual activities that show some analogy with such narrowly characterized TEs, and to this we now briefly turn. Let us distinguish two families of such activities: first, those that have as their primary purpose increase of knowledge like the strict TEs do, and those that have other purposes, but in their construction resemble strict TEs.

One very close area is linguistics. In the field research, the linguist presents a string of sounds (phonemes, letters) to the native speaker of the language investigated and asks her questions. Typically, she might ask the speaker to decide if such-and-such would be a sentence of his language. Or she might present a sentence with interesting use of pronouns, like "Tom thinks Dick loves himself". To whom does "himself" refer, to Tom or to Dick, the linguist asks. The native speaker might react immediately, or try to reply the sentence himself. The verdict will be an intuition, for instance, yes, this is a sentence of my language. Or, "himself" refers to Dick, not to Tom. The appeal to such intuitions has been playing a paramount role within Chomsky's paradigm of linguistic research. It is possible that the success of the paradigm has made intuitions and TE-like contexts in general more acceptable to analytic philosophers inspired by science, including linguistics and cognitive science.

But there is much more. Some hundred years ago, more precisely in 1905, Ernst Mach has proposed a list of various relatives of scientific and philosophical TEs, even describing them as TEs themselves:

> Besides physical experiments there are others that are extensively used at a higher intellectual level, namely thought experiments. The planner, the builder of castles in the air, the novelist, the author of social and technological utopias is experimenting with thoughts; so, too, is the hard-headed merchant, the serious inventor and the enquirer. All of them imagine conditions, and connect with them their expectations and surmise of certain consequences: they gain a thought experiment. However, while the former combine in fantasy certain conditions that never occur together in reality, or imagine these conditions accompanied by consequences that are not connected with them, the latter, whose ideas are good representations of the facts, will keep fairly close to reality in their thinking. Indeed, it is the more or less non-arbitrary representation of facts in our ideas that make thought experiments possible. (Mach 1976:29)

I propose that we follow Mach in opting for a very wide genus, but that we keep the name "thought experiments" for the narrower, more controlled, cognitively focused species (many author simply write about literary works as about TEs, the most prominent these days is Catherine Elgin (see, for instance her 2007 paper). We then need another name for the whole; I propose "imaginative enactments in thought", "IET" for short. Such enacting happens in theoretical contexts, in everyday understanding of others, in practical planning, maybe elsewhere in the vicinity of planning. Finally, there is fantasizing for pure pleasure, also a kind of imaginative enactment. Here we shall concentrate on kinds closest to TEs. The phenomenon has been noticed early, at the beginning of the twentieth century by E. Mach, and that he proposed keeping the term "thought experiment" for the widest range of activities:. There is a tendency on the present-day scene terminologically to follow Mach, and take "TE" in very wide sense. The recent *Routledge Companion to Thought Experiments* (Davies 2018) figures a chapter on "Art and Thought Experiments" by David Davies that does exactly that, and the list can be made longer. We shall here take talk about other TE-like activities of imaginary enactment, having a different primary motivation than strict TEs, as about the wider family or genus, and propose a name for it. So, let me then list some close relatives of TEs, and say a few words about each.

As Davies rightly notes, a very wide are of imaginative enactments is to be encountered in fiction, either literary, or theatrical (drama, opera, even ballet), cinematographic or even perhaps in painting (see, for instance, Geordie McComb's "Thought Experiment, Definition, and Literary Fiction" (McComb 2013). The primary function of the presentation might be artistic and/or enjoyment inducing, but normally there will be a cognitive point as well. Take the "Matrix"; it offers a scenario in which brains are manipulated by evil scientists, and induced to quasi-perceive and accept a completely illusory reality. It is very close to the famous TE of Evil Demon, due to Descartes, and even closer to its more recent version of the Brain-a-vat due to Putnam. In continental philosophy, from the second half of the twentieth century, the fictional genre is the dominant kind of TE-like items to which philosophers refer.

The cognitive aim of coming to know and to understand can be, and often is present there, but the requirements of strictness are weaker than in TEs. In science and philosohy the TE should have a clear and univocal goal, and the proposal that is tested by it has to be decided in a non-ambiguous way. In a literary work ambiguity is often praised as a goal. Take Ursula Le Guin's (1974) novel *The Dispossessed-an Ambiguous Utopia*, where fiction is used to create a mega TE, an interesting contrast emerges. It is not exactly the contrast between fiction and non-fiction, but between unambiguous and guided imaginative exercise versus ambiguous, open textured one. (For Le Guin's TEs see also Whittington, Zoe; Jacob, Smith; Donovan, Muniz (2020), "The Gender Thought Experiment in Ursula K Le Guin's *The Left Hand of Darkness*" at URI: https://scholarsbank.uoregon.edu/xmlui/handle/1794/25543).

The next species encompasses motivational works, utopias and dystopias. They are very similar to political TEs (say, Plato's *Republic*), but their main role is motivational one: to be producer of ideals, positive or negative (e.g. in the negative utopia) that should prompt action in a relatively direct fashion. Once the demand for implementation becomes central, the political TE (in a wide sense) might become a motivational and applied one; most often it moves in the direction of positive or negative utopia. Of course, the same text is sometimes read as more epistemic, or more motivational. The epistemic/motivational contrast is important because of different functional roles of the scenarios imagined, and their consequences. For instance, being unrealistic is not fatal for an epistemic TE and might even turn to its advantage, since the absence or realistic detail might enhance the clarity of the picture proposed. In contrast, being unrealistic is fatal, or at least seriously disabling for a motivational scenario. Further, mistakes in theoretical TEs have different sort of consequences than mistakes in motivational ones: unfortunately, implementing an ideal brings much more serious problems in its wake than just discussing it. A still actual and ironical example is offered by the history of communist ideal, developing from political TE (utopia), through "science" and practice, from Engels to Brezhnev so to speak, and collapsing in implementation.

The final kind we are going to mention is meditation, either of a therapeutic or of a religious kind. These exercises allow for much wider variation and topical freedom than do philosophical TEs. In therapeutic controlled meditations pluralism is expected: the subject can adapt the content to her own needs, and it is the result that counts. (TEs, in contrast, command a narrow focus: philosophy and science are special in raising high demands, here the cognitive upshot is the crucial motive, and it is characterized in demanding terms, including singularity of outcome. This results in the search for a unique, presumably right, answer to a precise question; they are highly controlled, rationality demands are at their highest and the role of emotional states is consciously minimized. The question is primarily theoretical, truth-oriented, not character-building or emotion-provoking.) So, once we consider other kinds of imaginative enacting in thought, and start comparing TEs to other IETs, the former seem to belong to a real kind.

Religious mediations, for instance the ones due to Loyola or saint Bonaventura also leave a wide space to the reader-meditator. In the case of religious meditations, we have an interesting link to philosophy. Here are Descartes's *Meditations on First Philosophy*, with the TE concerning an Evil demon:

> Accordingly, I will suppose not a supremely good God, the source of truth, but rather an evil genius, supremely powerful and clever, who has directed his entire effort at deceiving me. I will regard the heavens, the air, the earth, colors, shapes, sounds, and all external things as nothing but the bedeviling hoaxes of my dreams, with which he lays snares for my credulity. I will regard myself as not having hands, or eyes, or flesh, or blood, or any senses, but as nevertheless falsely believing that I possess all these things. Descartes (2006:12)

And then comes the analogy with real meditations:

> I will remain resolute and steadfast in this meditation, (manebo obstinate in hac meditatione defixus), and even if it is not within my power to know anything true, it certainly is within my power to take care resolutely to withhold my assent to what is false, lest this deceiver, however powerful, however clever he may be, have any effect on me. But this undertaking is arduous, and a certain laziness brings me back to my customary way of living (Descartes 2006, p. 23).

Indeed, in answering the objections of his critics, Descartes counsels that we pass long time, say a few months or weeks, concentrated on imagining the Evil Demon situation; such meditative effort will make things clearer to us: "And I would like my readers not just to spend the short amount of time it takes to get through the Meditation but rather to devote a few months or at least a few weeks to the topics requiring consideration before going on to the rest of the work. For in so doing, they no doubt could gain much greater profit from the rest of the work." (Descartes 2006:76) (for a fine comments see Rico Vitz (2015)).

We thus have several kinds of imaginative activity close to thought experimenting, and others might be in the offing. Obviously, TEs are a species of a more general genus, that encompasses examples from fiction, from linguistics, from political utopianism and from meditative activities. Let us call this wider genus, or family, "imaginative exercises in thought", IETs for short. Here is the summary of what we have been discussing:

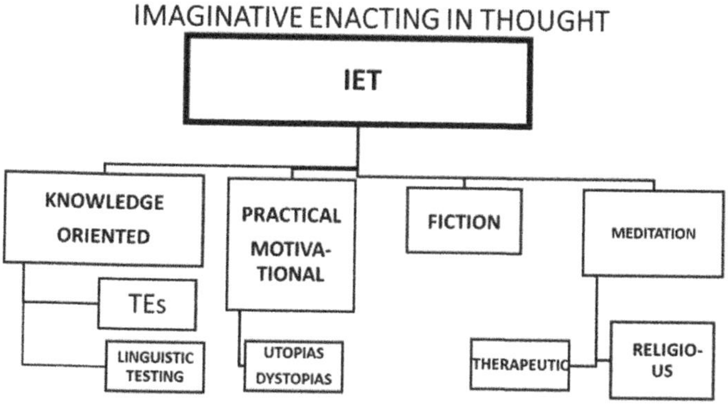

Some authors add a third kind of scientific testing, namely computer simulation. It shares with the laboratory experiment the exactness and possible wealth of imputes, and with scientific TEs the non-material kind of data processing.

Here is then the other division: We thus have two trees. The first connects TEs with other, not scientific imaginative activities grouped together under the name of ITS. The second connects them with scientific testing procedure, picking up scientific TEs as a relevant sub-species. If we combine the two trees we get a perspicuous representation of the place of TEs in the scheme of human intellectual activities:

All in all, variations are very wide reaching. Let me note similar variations in the everyday use of the term "experiment." The *Meriam-Webster Dictionary* which offers as its first example a few sentences that depicts scientific practice, talking about "simple laboratory experiments" and "some experiments with magnets." But then it passes to "an experiment in living more frugally" and concludes with "the city's experiment with a longer school year." These last two are not the province of philosophy of science for sure; and we have no problem to distinguish full scientific meaning of "experiment" from its less demanding relatives. Introducing the wide notion of IETs solves the problem of vagueness, although the tendency in the recent literature is to group many typical kinds of IETs under the (wide) rubric of "thought experiments". We prefer to keep the notion of thought experiment more precise, and to keep their relatives (in linguistics, fiction and so on) under distinct names that they traditionally bear. The notion of IET does sufficiently point to the similarities, and nicely organizes the whole family.

1.5 Conclusion

Thought experiments (TEs for short) have been for centuries, if not for millennia, useful and in some areas indispensable devices for scientific and philosophical theorizing.

In this chapter we have recounted a number of famous TEs from science and philosophy and offered a tentative characterization of TEs in general. They are similar to laboratory experiments. A typical TE starts with a design, which involves the determination of the goal(s), in the thesis/theory to be tested, and the construction of a scenario to be considered.

It then proceeds with the presentation of the scenario thus constructed to the experimental subject, either the author of the scenario, or an interlocutor. In the later situation, the testing is done independently of the author: she is supposed to sit and wait for the verdict from the interlocutors, i.e. experimental subject's 'laboratory of the mind'.

On the side of the subject, the experiment start with understanding of the proposed scenario, and then continues with the typically imaginative contemplation of it. Some reasoning might intervene. If all goes well, the subject ends with a verdict concerning the thesis/theory to be tested. Usually, in her mind it is presented to her as an invitation to believe or directly as a belief, most often seeming obvious and compelling. Such states (invitations to believe, or immediate beliefs) have been traditionally described as "intuitions"; they are often likened to similar states concerning mathematical insights or obviously looking linguistic judgments on sentences in subject's native language.

Once the verdict is achieved, it can be and often is compared with results of other scenarios in the vicinity, or other versions of roughly the same scenario. Finally, interesting and provocative verdicts are normally being brought to comparison with items of knowledge or widely accepted beliefs. If they clash, the arduous task of balancing is required, in which the particular verdict might win (as has historically been the case with Galileo's verdict on falling bodies), or, alternatively, the established knowledge might, or, thirdly, some compromise is made. The result is usually described as "reflective equilibrium".

Recently, in addition to being widely used, TEs have themselves become a hot topic of debate. In philosophy of science and general philosophical methodology (or meta-philosophy, to use a fashionable term) one debates their characteristics and definitions, their role, potential limitations and ways to overcome them. I would add that a big part of history of analytic philosophy, and some parts of general history of philosophy, can be understood as consisting of traditions starting with particular thought experiments, and then proceeding with variation and prolonged argumentation until the present moment, and promising to continue in the future; we dedicate Chapter Five to this phenomenon.

Finally, following Ernst Mach, we have noted that TEs are a species of a wide, interesting and important genus, encompassing other, somewhat related imaginative exercises in thought, like the ones in linguistics, in fiction, in practical political imaginings and in meditation. On the side of science, the scientific TEs have important counterparts in laboratory experiments, and, in recent times, in scientific computer simulations.

References

Brown JR (1991) Laboratory of the mind: thought experiments in the natural sciences. Routledge, London

Brown JR (1999) Philosophy of mathematics: an introduction to the world of proofs and pictures. Routledge, London

Davies D (2018) Art and thought experiments. In: Stuart M, Fehige Y, Brown JR (eds) Routledge companion to thought experiments. Routledge, London, pp 512–525

Descartes R (2006) Meditations, objections, and replies, edited and translated by Ariew R, Cress D. Indianapolis: Hackett Publishing Company, Inc., Cambridge

Einstein A, Infeld L (1938) Evolution of physics. Cambridge University Press, Cambridge

Freeman SJ (2007) Rawls. Routledge, London

Gettier E (1963) Is justified true belief knowledge? Analysis 23:121–123

Le Guin UK (1974) The dispossessed-an ambiguous utopia. Harper and Row, New York

Häggqvist S (2009) A model for thought experiments. Can J Philos 39(1):55–76

Hilbert D (2013) On the infinite. In: Ewald D, Sieg W (eds) David Hilbert's lectures on the foundations of arithmetic and logic 1917–1933. Springer,Heidelberg

Mach E (1976) Knowledge and error. Reidel, Dordrecht

McComb G (2013) Thought experiment, definition, and literary fiction. In: Frappier M, Meynell L, Brown JR (eds) Thought experiments in philosophy, science, and the arts. Routledge, London, pp 207–221

Norton JD (2013) Chasing the light: Einstein's most famous thought experiment. In: Frappier M, Meynell L, Brown JR (eds) Thought experiments in philosophy, science, and the arts. Routledge, New York and London

Rawls J (1999) A theory of justice, revised edition. The Belknap Press of Harvard University Press, Cambridge, Massachusetts

Rawls J (2001) Justice as fairness: a restatement, Kelly E (ed).Harvard University Press, Cambridge, MA

Roux S (2011) Introduction: the emergence of the notion of thought experiments. In: Ierodiakonou K, Roux S (eds) Thought experiments in methodological and historical contexts, Leiden and Boston

Searle J (1980) Minds, brains and programs. Behav Brain Sci 3:417–457

Searle J (1999) The Chinese room. In: Wilson RA, Keil F (eds) The MIT encyclopedia of the cognitive sciences. MIT Press, Cambridge, MA

Gendler TS (2004) Thought experiments rethought—and reperceived. Philos Sci 71:1152–1163

Vitz R (2015) Reforming the art of living-nature, virtue, and religion in Descartes's epistemology. Springer, New York

Chapter 2
The Life and Importance of TEs

2.1 The Stages of TE

Let us now turn to a description of the ways TEs normally function. We shall first consider a typical TE and its 'biography', the stages it normally goes through. Then, we shall briefly address the question of the point of thought-experimenting: what is the role of TEs in science, mathematics and philosophy? Next we shall briefly discuss the immediate result of a TE, usually described as "intuition" generated by the TE. Finally, we shall turn to the role and importance of TEs in science and philosophy.

We have characterized a TE is a process that starts with a design, which involves the determination of the goal(s), in particular the thesis/theory to be tested, and the construction of a scenario to be considered. We noted that it then proceeds with the presentation of the scenario thus constructed to the experimental subjects. On the side of the subject, the experiment then continues with the typically imaginative contemplation of the scenario plus some piece of reasoning, culminating in the decision ("intuition") concerning the thesis/theory to be tested. Once this result is achieved, it can be, and often is compared with results of other experiments in the vicinity. So, one topic that seems natural as a starting point in studying TEs themselves, namely characterization of stages they go through. Surprisingly, philosophers haven't done it systematically; so let us consider the stages now. I shall follow our academic practice of using TEs and write as if a TE is normally invented by someone, and then presented to an audience, to be called "the subject" in the sequel.

Here is the list of stages I would propose, illustrating it with reference to Galileo's famous one:

We can think of a preliminary stage zero, preceding the experimenting itself, in which the creator of the TE invents the design of the experiment. We than pass to the stages in which the creator, enters the dialogue with the listener.

At stage one, the creator of the TE presents the scenario imagined and asks the question. In Galileo's TE the question seems simple: Would the body composed of a cannon ball and a musket ball fall faster or slower than the light one (the musket ball) and the heavy one (the canon ball)?

N. Miscevic, *Thought Experiments*,
https://doi.org/10.1007/978-3-030-81082-5_2

If everything goes well, at the stage two the interlocutor arrives at understanding; for example, in Galileo's dialogue Simplicio gets the question right. The task might seem trivial, but a lot of discussion of TEs, especially these days, turns around the ways subjects understand the presentation and questions (see Chap. 5 for more about it).

At the stage three on the side of the interlocutor one has tentative conscious production. In the example, we assume that Simplicio tries to imagine the arrangement imagine the cannon and the musket ball falling and, using naïve physics, does it to his satisfaction.

Here we must be cautious and allow for a stage four, a possible non-conscious elaboration. In our example, Simplicio might unconsciously doubt that the composite thing is an individual body at all, or he might ask himself what would happen if the two bodies were of very different kind, say a cannon ball and a sponge tied together.

At stage five, we have the immediate spontaneous answer, the seeming and belief usually called "intuition". Simplicio has a clear intuition that the arrangement will not work, since the composite body would be falling both faster and slower than the canon ball alone.

At stage six one may proceed to varying and generalizing, as we often do in the class when we present a TE. In our example the main question is if the reasoning can be applied to any pair of objects. Is there anything special in the first example? What about cannon ball and sponge tied together? What if we tie a small cork to a heavy stone? A book and a leaf of paper glued together. (Historically, this has been happening to Galileo himself: the young Galileo was asking the question only about bodies of equal density; later he generalized the point to any two bodies). The procedure has been called "intuitive induction", and it is quite important in the life of a TE.

If everything goes well with intuitive induction, at stage seven the cognizer ends with a general belief. Simplicio might realize that the problem generalizes to every combination of falling bodies and that in every case his commonsense "theory" gives an impossible prediction.

Here is an overview:

		Galileo's TE
Stage 1	The question	Salviati: *Would the body composed of a cannon ball and a musket ball fall faster or slower than the light(musket) and the heavy (canon) one?*
Stage 2	Understanding: *Simplicio gets the question right*	*Question understood*
Stage 3	Tentative conscious production: *Simplicio tries to imagine the arrangement and does it to his satisfaction*	*Trying to imagine the cannon and the musket ball falling (using naïve physics)*
Stage 4	Possible non-conscious elaboration	possible non-conscious elaboration

(continued)

(continued)

		Galileo's TE
Stage 5	Immediate spontaneous answer *The arrangement will not work...*	*This composite body would be falling both faster and slower than the canon ball alone*
Stage 6	Varying and generalizing *Can it be applied to any pair of objects?*	*Is there anything special in the first example? What if we tie a small cork to a heavy stone?*
Stage 7	General belief *In every case my theory gives an impossible prediction*	*The problem generalizes to every combination of falling bodies*

Let me now point to yet another example, a famous twentieth century TE from physics, Heisenberg's gamma-ray microscope one, described in his famous paper from Heisenberg (1927). Here, the issue is whether position and momentum of a small particle, say an electron, can be measured at the same time; Heisenberg will, of course defend the negative answer, in line with the indeterminacy principle for elementary particles. The basic idea is extremely simple and elegant. Suppose you want to measure the position and velocity of an electron and you do it by scattering gamma rays, as in the original version of Heisenberg (1983: 4) or photons off it; start by one item of each. Now, the lower the wavelength the higher the resolving power of a microscope, so that only high-energy electromagnetic radiation has a short enough wavelength to detect an electron within any reasonable range of error. Therefore, you have to use high-energy gamma rays (or photons) since there is an inverse relationship between wavelength and energy. The powerful gamma-ray will give you the exact momentum of the electron, but its position will stay indeterminate. Inversely, if you want the exact position, you have to use low energy ray, but this one will have a wide diffraction pattern, so no way to detect the exact position of your target, since the high energy gamma-ray will push the electron around, so you will not be able exactly to determine its position.[1] Now, the impossibility might just be an accident, so Heisenberg has to show that it will persist no matter how circumstance change. So, the first objection, aimed at the photon example might be that this is just an imprecision of optical instruments (see Frappier 2004: 25). Heisenberg can then vary, and introduce different particles; we already had gamma-ray in our presentation. The result is the same, so the first variation is successful for Heisenberg's proposal.

But what if elementary items are not particles at all, just waves, to don't have a position in the sense in which particles would have one. No, even if our item is a wave, it has a position, so the argument holds.

Third, momentum might be indeterminate, but position is not. Momentum affects the position measurement, but not vice versa. So, measure position first, and then turn to momentum. Heisenberg considers using Doppler Effect for measurement, but concludes that this will not change the general situation.

---

[1] I am following Kent A. Peacock's short overview (Peacock 2018) and Melanie Frappier's (2004) extremely interesting dissertation on Microscope TE.

This is not the end of the story, but it suffices to illustrate the strategy of variation. Interestingly, Frappier talks of induction in these cases (Frappier 2004:25 ff.), but we should be aware that it is "intuitive induction", reasoning from imaginarily varied imaginary options.

Now, the same variational structure characterizes mathematical TEs. Consider Hilbert's hotel. Hilbert invites the reader to assume that the hotel has infinitely many rooms numbered 1, 2, 3, 4, 5, … and that each of the rooms is occupied by a single guest. So, the hotel is, so to speak, full.

Now, a new guest arrives. So, imagine, he suggests, that the manager moves each guest into the next room. Then comes the crucial question: is there going to be a room available for the new one? The scenario is easily understood by the reader. Then, at stage three follows the contemplation and enactment of the scenario. If all goes well, at stage five we have a positive intuitions: Oh yes, there will be room for the next guess. Now, we can vary the story a bit: what if the situation is repeated with another guest arriving? And the general conclusion is that adding guests one by one does not change the availability of rooms. Infinity is richer than the reader might have initially expected.

The same structure can be detected in philosophical TEs. Let me mention a topic in practical philosophy. Here is the famous "Trolley problem" due to Philippa Foot and formulated in 1967. Here is a simple formulation due to Judith Jarvis Thompson, who made a significant contribution to the discussion of the TE:

> Suppose you are the driver of a trolley. The trolley rounds a bend, and there come into view ahead five track workmen, who have been repairing the track. The track goes through a bit of a valley at that point, and the sides are steep, so you must stop the trolley if you are to avoid running the five men down. You step on the brakes, but alas they don't work. Now you suddenly see a spur of track leading off to the right. You can turn the trolley onto it, and thus save the five men on the straight track ahead. Unfortunately,...there is one track workman on that spur of track. He can no more get off the track in time than the five can, so you will kill him if you turn the trolley onto him (Thompson 1985: 1395).

Here are the stages: We begin with stage one, the question being asked. Stage two, the question is understood by the subject. Stage three offers the tentative conscious production, say building the picture of the two tracks with workers, all done at conscious level. Stage three brings unconscious additional production and probably controlled by the relevant competence at unconscious level (some geometry and commonsense physics might be needed to imagine the scenario in sufficient detail). At stage four the subject arrives at immediate, spontaneous verdict, often non-conscious, for instance "Yes, I would turn the lever and save the five men". One might think of an additional stage of sub-personal empirical theorizing by Central Processing Unit. (I, the reader, might imagine workers from abroad, say Mexicans, I might imagine young and healthy workers, or older and tired ones, and so on, all motivated by my views on the working class and the like). At stage 5 comes the immediate spontaneous answer (= intuition): "Yes, I would turn the lever and save the five men". At stage six we have varying and generalizing, intuitive induction at both conscious and unconscious levels. For instance, it seems to me in the trolley case, that I would turn the lever and thus save five by sacrificing two; but what if the two are (a) small children, (b) very talented artists, (c) my friends? Stage seven offers the general belief, for instance that I would turn the lever no matter what.

We might think of a further stage in which I wonder how the result, the general belief, fits with my other considered judgments (intuitions), with theories I believe in and so on? For instance, I would turn the lever since I think five lives are more valuable than two, no matter whose lives the latter are. But why do you believe this? Why don't you give additional weight to children, since they have more time left to enjoy their lives? Because I think the value of each life is the same as of any other. And so on. If I arrive at a satisfactory view of the whole, this will yield what is called a "reflective equilibrium" at conscious level in which my views from an equilibrated structure.

Why is it important to pay attention to the stages? First, for the understanding of actual TEs, as performed in the armchair, classroom seminar and the like. Different misunderstandings occur at different stages. A student might verbally misunderstood the stora and the question. Or she might wonder why the question is asked at all. This is very different from, say, doubting the relevance of the particular example. Second, the understanding of the longer history of each given TE in the history of philosophy or of physcis, might profit from the analysis of stages; for instance Evil demon or Social contract. These are long term histories, each lasting around three hundred years if we count only modern philosophy. If we could bring together short-term and long-term stages, this would make the historical process more intelligible. Social contract has generated the Original position proposal. What is the relation between such historical changes, and the usual micro-variations of a given TE? Once we have a worked out the answer we shall be able to integrate the meta-philosophy of TEs with information from history of philosophy (for more, see Chap. 6).

Third, the attention to stages is important for making our thought experimenting better. Skeptics about TEs have been challenging thought-experimenters with lots of empirical material, which shows that subjects are often unreliable, and often open to external, completely irrelevant influences. Jonathan Weinberg has, for instance, argued in 2007 (see References), that intuitions lack all possible sources of 'epistemic hopefulness': no external corroboration, no internal coherence, no testability of margins of reliability and no theoretical illumination! (Weinberg 2007, pp. 327–330). Conclusion: intuition is hopeless as a source of evidence! Capellen has, in his (Capellen 2012) book, replied with the claim that philosophers don't take TEs seriously anyway! Now attention to stages would help us recognize dangers, and distinguish them in regard to their origin; some might come from pragmatic sources, from misunderstanding to the desire to please, others from wrong-headed generalizations, or lack of attention to the right variety of cases from which to generalize, and still other might come from deep defects in our cognitive system. This might help with the ameliorative project. Daniel Nolan recently wrote that "/…/you might expect that there would be self-help works available for improving one's faculty of philosophical intuition, and exercises to improve intuition reporting. There is little in the philosophical literature aimed at this, however." (Nolan 2016, p. 163 ff.). We badly need such works geared to performing an ameliorative function, and attention to stages might be an important step. We shall return to it in the Chap. 4.

Let me conclude the section by adding two famous philosophical TEs; I leave the identification of possible stages of performing each of them as an exercise for the reader.

First, a TE due to Lehrer (1990:163–164), supposed to show that reliable information is not sufficient for knowledge. Imagine that an experimental surgeon invented a thermometer-and-a computational small device capable of generating thoughts. The device is then implanted in the head of a person, called "Truetemp"; it reliably transmits correct information about the temperature to the computational system of his brain. Now, poor Truetemp has no idea that the device has been inserted in his brain, has thoughts about temperature all the time, but never checks a thermometer to determine whether these thoughts about the temperature are correct. Thus, he thinks and accepts that the temperature is, say, 104°, Lehrer says. And his thought is true and due to the reliable source in his head. But does he know that the temperature is 104°? We shall return to the TE briefly in the last chapter.

Our next example is a famous TE due to Gettier (1963:123). We are told that Smith has strong evidence for the following proposition:

(f) Jones owns a Ford.

Smith's evidence might be that Jones has at all times in the past within Smith's memory owned a car, and so on. Imagine, now, that Smith has another friend, Brown, of whose whereabouts he is ignorant. He selects three place names quite at random and constructs the following three propositions:

(g) Either Jones owns a Ford, or Brown is in Boston.
(h) Either Jones owns a Ford, or Brown is in Barcelona.
(i) Either Jones owns a Ford, or Brown is in Brest-Litovsk.

Gettier reminds us that (Ig), (h) and (e) are entailed by (f) Jones owns a Ford. Smith realizes that this is so, and accept all three of them ((g), (h), and (i)) on the basis of the proposition that Jones owns a Ford. Smith has correctly inferred the three from a proposition for which he has strong evidence, Gettier reminds us. Smith is therefore completely justified in believing each of these three propositions. And then he introduces two additional conditions. First, Jones does not own a Ford, the car he is driving is a rented one. And secondly, by sheerest coincidence, and entirely unknown to Smith, Brown is in fact in Barcelona. So, it is true that (h) Either Jones owns a Ford, or Brown is in Barcelona. Smith believes that it is so, and is justified in his belief.

Still, and this is the point, Smith does not *know* that (h) is true. So, justified true belief is not sufficient for knowledge.

Again, we shall return to the example in later chapters.

2.2 Intuition

In the discussion of TEs, most of the attention has been focused upon the latest of the early stages of a TE, namely the intuition that a TE is supposed to result in. Note that the reader of Hilbert's text is supposed to 'see' that there will always be a room available to the new guest, the reader of the Trolley problem proposal to see that turning the lever would be the right (or allowed, or prohibited) thing to do. The solution of the problem raised by the scenario, i.e. the answer to the leading question of a TE is presumably 'seen', and this is just due to the peculiarity of the medium. The 'seeing' is usually described as "intuition". What is this 'seeing'? Where does it come from? These are the central descriptive-explanatory issues that guide theoreticians who discussed TE in further matters of justification of people's thought-experimental, intuitional beliefs.

Let us briefly discuss the term and the concept of "intuition". The prime use of the term "intuition" is for the clear seeming and/or conviction, like the convictions from our two examples, Hilbert's room and Trolley problem. The meaning of the word "intuition" is tied to obviousness and immediacy. Indeed, the original philosophical meaning of the word has been applied to the cases of a clear insight, "intuitio" meaning simply "seeing". (The word appears in the middle ages, in optics, in theories about immediate perception and in the religious context of "seeing God".) Descartes has made it famous, linking it to "clear and distinct cognition" in his "Regulae" and "Principles". He speaks of that

> which is present and apparent to an attentive mind, in the same way as we assert that we see objects clearly when, being present to the regarding eye, they operate upon it with sufficient strength (Principle XLV, 1985, p. 207).

A slightly different way of understanding the term "intuition" comes from the Kantian tradition, and has also been prominent in the writings of German mathematicians, and others who have followed their usage. It focuses upon quasi-perceptual awareness, of the kind illustrated above with folding of a square. You can normally "see" the result in your imagination, and this kind of exercise is often categorized under the rubric of "intuitive geometry". Kant suggested in his *Logic* lectures the Latin term "intuition" as a translation of his "Anschauung", encompassing both perception and quasi-perceptual imagination. Hilbert talks about "intuitive geometry" meaning exercises like the one above.[2] So, the states we call "intuitions", vary in regard to their topic, but they share their phenomenal properties classically described by Descartes in terms of "clear and distinct cognition", of being "present and apparent to an attentive mind". L. J. Cohen proposes that "...an intuition that p is just an immediate and untutored inclination, without evidence or inference, to judge that p" (Cohen 1981, p. 318).

---

[2] His book *Anschauliche Geometrie* (1932) has been translated in 1952 as *Geometry and imagination*, but the tradition of using "intuitive" for "quasi-perceptual" has stayed with philosophers of mathematics and science. Poincaré in his *The Value of science* writes "Logic (...) is the instrument of demonstration; intuition is the instrument of invention ..." (Poincaré 1958:23).

Similarly, the S. Blackburn's *Oxford dictionary of philosophy*, speaks of "immediate awareness, either of the truth of some proposition, or of some object of apprehension such as a concept" (Blackburn 1994, p. 197). Michael Ayers—commenting Locke, and apparently endorsing his general stance–describes the intuition of necessity in the following way:

> ..the perception of necessity is something like seeing, but is more like ordinary cases of the intelligent apprehension of the significance of what we see, such as the realization how this (actual or depicted) gear works (Ayers 1991, p. 299)

We shall follow the mainstream in distinguishing an intuitional seeming, and the final, settled spontaneous intuitional belief. Seeming has a degree of compellingness, Sosa talks about attraction, and characterizes intuitional belief in terms of assent to attraction. (Since for Sosa it is all at the animal level, the beauty and the beast come to one's mind: the beautiful intuition-content attracts the animal attention and belief-forming capacity, and the later gives its assent).

Now we have to introduce a few distinctions concerning various uses of the word "intuition". We have taken it to refer primarily to cognitive states (those of 'seeing' that some proposition holds). It also denotes the contents of intuitive judgment (E.g. "John has this strange intuition that eating vegetables is morally wrong."). Moreover, philosophers have hypothesized that there is a capacity for having intuitions, and have extended the use to the term to cover the capacity. So the term "intuition" has at least three meanings:

(1) intuition-state, i.e. either inclination to judge that p (with the feeling of obviousness and irresistibility), or the judging itself,
(2) intuition-content, i.e. the judged proposition that p,
(3) intuition-capacity, i.e. the hypothetical capacity producing and underlying the intuition-state.

Add to these standard meanings the process producing intuition, that is occasionally described as "intuiting that p".

Intuition state		Intuition-content	
State specification	State-features	Modal specification	Core content
I believe I "see" I intuit	Certainly Clearly Obviously/that Indubitably-compellingly	Necessarily	*Continuity determines identity of material objects* $2+5=7$ *if* p and q *then* p

Here we shall assume the view that there are intuitions-dispositions and judgments, which form a distinct group of phenomena, and there is the intuition-capacity, the capacity to use our imaginative and judgmental competencies in an off-line fashion. It is the voice of competence, say the one responsible for naive physics, and then spatial-geometrical, metaphysical, moral, linguistic and other competencies.

The next thing to note is the primacy of the concrete. TEs and the intuitions they produce are rather scenario-based than inference-based. Imagining scenarios, typically particularized ones play the main cognitive and justificatory role, whereas inference typically plays a subordinate role.

One issue that has been prominent in the discussion is the contrast between 'extroversion' and 'introversion': are intuitions concerned with their external objects, the domain of items and facts, or with our concepts? Is Galileo investigating the properties of falling bodies, or of the concept of the falling body? My sympathies are with external reference. Concepts often play a role in the process, but they are not the object of intuitions, and their role is subordinate to the role played by the external referential domain.

Of course, intuitions require an explanation of having and reliability, if possible a causal one. The heart of philosophy is the question "how possibly" (as famously noted by arch-apriorist Kant), and in typical descriptive matters the question is naturally read as demanding a general causal account of the phenomenon in question. My own favorite line is summarized in the following paraphrase of J. L. Austin (where is term "words" is replaced with "cognitive inclinations"): the stock of our deepest cognitive inclinations embodies all the distinctions men have found worth drawing, and the connections they have found worth making, in the lifetimes of many generations: these surely are likely to be more numerous, more sound, since they have stood up to the long test of the survival of the fittest, and more subtle, at least in all ordinary and reasonably practical matters, than any that you or I are likely to think up on the spot (Austin 1980, p. 182). This accumulated wisdom then allows the philosopher to anticipate the experience from the armchair. At the same time, the double fallibility of these intuitions accounts for the limits of philosophical autonomy: armchair research should be open to corrections from empirical science.

Let me finally mention a contrast that dominates epistemological discussion of intuitions, the one of a priori justification versus the a posteriori one. A priori justification is a type of epistemic justification that is, in some sense, independent of experience, Stanford Encyclopedia tells us (see "A Priori Justification and Knowledge"). And the article goes on to list intuition as one of the main source of a priori justified beliefs. Indeed, historically, the main candidate for a priori knowledge are beliefs derived from and based upon reason in the wide sense, that encompasses intuitions as well. Descartes had divided them into immediate and mediate ones: the former are intuitive insights, and the later are derived from the former by inference and reasoning. Traditionally, intuitions are seen as being justified a priori.[3]

We shall discuss the issue of justification of intuitional beliefs in Chap. 4; here let me just say that I believe that most of them carry both types of justification, in a structured and determinate manner. For example, Stevin's TE and the resulting intuition that the chain will not move, deploys some spatial-geometrical knowledge that might be innate and in this sense a priori, some naïve physics that is partly innate

[3] The traditional apriorist view has been defended by authors like BonJour, (1998) and Casullo (2003, 2012); for further reading see Boghossian and Peacocke (eds.) (2000). For the empiricist alternative see Devitt (2005). For an interesting third way proposal see Williamson (2013).

(a priori) and partly derived from and justified by experience (a posteriori), and we can trace each of the lines of justification to its distinctive source.

2.3 The Role of TEs

Why do scientists, mathematicians and philosophers engage in thought-experimenting? What is the goal of all this? In science and mathematics TEs clearly have a heuristic value. Thomas Kuhn has persuasively argued that TEs help us to re-conceptualize the implicit knowledge we have about the world. He has drawn an impressive parallel between this role of TEs and the function of scientific revolution.

> What was vaguely known in spite of the community's mental equipment before the revolution is afterwards precisely known because of its mental equipment. (...) Conversely, thought experiment is one of the essential analytic tools which are deployed during crisis and which then help to promote basic conceptual reform. The outcome of thought experiments can be the same as that of scientific revolutions: they can enable the scientist to use as an integral part of his knowledge what that knowledge had previously made inaccessible to him. That is the sense in which they change his knowledge of the world (Kuhn 1977, p. 264).

Other authors stress the role of singular imagining in a TE, in contrast with general argumentation; a TE, according to them, "depicts particulars in ways that make manifest practical knowledge, and describes an imaginary scenario wherein relevant features can be separated from those that are inessential to the question at issue." This is what Tamar Gendler writes commenting Galileo's free fall TE (Gendler 1998, p. 420). She finds TEs indispensable for science.

TEs can thus teach us about directions in which to think about deep issues in science. Sometimes they bring us new knowledge. But many authors like to stress a specific contribution to understanding that they can produce. Stuart (2016:25 ff, and 2017) stresses the fact that they can make the reader see meaningful connections between the relevant concepts, and even to think of new applications; he sees these contributions as essential to understanding.

It seems that the role of TEs in philosophy is even bigger than in science. Philosophy does not use laboratory to test its theories; the only experiments available here are those in thought. *TEs play in philosophy the crucial role that laboratory experiments play in science.* Take practical philosophy: the traditional sources of insight here are either facts (including presumed facts), principles or TEs. Facts are useful and indispensable, but taken alone they don't teach us about what is valuable, morally prohibited, morally indifferent and so on. We need at least principles. But how do we test principles? The only source here are intuitions and the indispensable testing grounds are TEs.[4]

Let us look more closely. We are interested in all sorts of cases, and we want to test our general principles: do they yield acceptable verdicts in particular cases? For this

---

[4] Thanks go to my colleague and friend Friderik Klampfer, for insisting on TEs being superfluous for ethics, see his Klampfer (2017).

testing, we need an independent spontaneous judgment (not necessarily a fast one), but singular and spontaneous for sure. How do I decide? First, by considering my own spontaneous reaction and then asking other people about their singular judgments. Here is Gilbert Harman, who is certainly not sentimental about TEs: "in both science and ethics, general principle are invoked to explain particular cases and, therefore, in both science and ethics, the general principles you accept can be tested by appealing to particular judgments that certain things are right or wrong, just or unjust, and so forth; and these judgments are analogous to direct perceptual judgments about facts." (Harman 1977, p. 6). Intuitions are like observations, pieces of evidence, indispensable in principle for moral(-political) thinking. If we abandon our appeals to spontaneous singular judgments, as pieces of evidence, we end in moral nihilism, and political Machiavellism of a very bad kind.

The next question is whether in practice we can rest satisfied with judging real cases only. Were we hyper-informed, about billions of real cases (each narrow case type with hundreds of variations), we might try to do with these, but we are not. One might criticize the Trolley TE by arguing that such cases are completely artificial, and that such choices never occur in real life; well, they do occur, for instance, in the time of war when the issue of sacrificing human lives is a burning one. But even in the war you don't have the number and systematicity of variations normal in the Trolley problem: what if it is five versus three, ten versus three, hundred versus three, what if one of the three is a genius that could help his people, or the world, what if… Just try teaching ethics using only actual historical examples! Real cases are simply not enough, neither to satisfy our moral curiosity, not to offer wide enough bases for generalizations TEs are needed as source of intuitions. We depend on imagined situations, and thereby on TEs to build, test and develop moral theories. Here, we simply cannot do without intuitions and TEs.

But isn't this just asking for a spontaneous judgment intuition testing? Without a concrete, particular example of a brother and a sister, as typically done in a TE, but still? No alternative to intuitional judgment is being offered! Frohlich and Oppenheimer propose something that looks like group thought experimenting, indeed with Veil of ignorance:

> In our experiments, groups of five people were subjected to conditions approximating a veil of ignorance. They were charged with the task of choosing, by unanimous, secret ballot, a rule for distributing income in a future society which they would inhabit, without knowledge of what role they would play in that society. (…) Groups were presented with a slate of possible rules, including those Rawls considered, but could introduce one of their own, or decline to choose a rule (2007, p. 128).

But is this really a deeply different alternative? No, it is simply an ethical TE.

The same goes for epistemology. The questions of whether and when one knows also contain a normative-evaluative component, and there is no wonder that TEs have been so prominent in epistemology, with Gettier TE as the showcase example. Remember, Smith has a true belief that the person with ten cents in one's pocket will get a job. But he thinks of the wrong person, Jones, and the information he is relying upon is in fact a misinformation (his boss told him that Jones will get the job). But he is innocent in trusting his boss, and good at deriving inferences from what he heard.

Plus, he knows that Jones has ten cents in his pocket. Does he know? The consensus is that he doesn't. And this TE has had revolutionary effects of twentieth century epistemology. TEs are indispensable in epistemology, it seems.

What about other domains of philosophy? Let me now mention another example of TEs from philosophy to the way in which they work in metaphysics and philosophy of mind. Let us turn to freedom and responsibility. People normally think that I am responsible for my action only if I could have done otherwise. If I cause a bad accident because it was completely impossible for me to act otherwise, I am blameless. In his "Alternate Possibilities and Moral Responsibility", Frankfurt (1969) offered a brilliant imaginary counterexample to this way of thinking. It involves two persons, a manipulative guy Black, and his poor victim Jones. Black very much wants Jones to act in a certain way, let us say perform an action of the kind A. He does not force Jones, however, although he has means for forcing him. He waits quietly "until Jones is about to make up his mind what to do, and he does nothing unless it is clear to him (Black is an excellent judge of such things) that Jones is going to decide to do something *other* than what he wants him to do" (p. 835). If it becomes clear that Jones is not going to do A by his own initiative, Black takes steps to ensure Jones does A. Black is the final winner, no matter how Jones decides. Question: if Jones does A without Black's intervention, is he responsible for doing it?

Here, the design of the experiment is reasonably simple, and should be clear to a non-expert reader-subject. The subject goes through the story, probably imagines the scenario herself, and rests satisfied with her imagining; there is nothing deeply problematic about it. Then comes the question. Well, Jones did A 'of his own free will', she will probably feel. "Yes, Jones is free in his doing", her impression, or intuition, will be. She will probably feel quite certain of it. This is the verdict of the laboratory of her mind. Now, in the context of philosophical investigation, say, in a seminar, she will be reminded of her other commonsensical conviction that one is responsible for one's action only if one could have done otherwise. We now have a conflict, and the need to balance. Frankfurt's paper offers a famous way out, a new reflective equilibrium, that our subject might appreciate as the solution, or reject, as being counter-intuitive.

Why is such a TE indispensable? Because philosophers are vitally interested in *connections between our spontaneous understanding of important properties, like freedom, and the results of science.* In order to answer the question about the relation between, say, scientific determinism and our belief in freedom, we need to confront them, and we cannot do it within science alone. We need the bridge, and TE is a perfect candidate.

Similar conclusion can be drawn about another central problem from the area. Does neurology explain our experience, in particular items like pain, experience of color and the like, sometimes called qualia? We need some account of our deep understanding of qualia. Well, we imagine a super-scientist, call her Mary, who had never had a particular qualitive experience, say, of seeing red (she lives in a black and white surrounding), but knows all there is to know about neurology. And then, one day, she sees a red surface. Will the experience be new to her? Yes, many readers are inclined to think. But then, neurology does not give you the knowledge of qualitative experience; there is a gap we don't know how to bridge. This is, of course, the famous Black-and-white Mary TE, due to Frank Jackson.

In short, TEs play in philosophy the role analogous to laboratory experiments in science: they are our indispensable guides and providers of understanding. Of course, not all philosophers agree, and we shall devote the whole Chap. 6 to disagreements.

References

Austin JL (1980) A plea for excuses. In: Philosophical papers. Clarendon Press, Oxford, pp 175–204
Ayers M (1991) Locke: epistemology and ontology. Routledge, London
Blackburn S (1994) Oxford dictionary of philosophy. Oxford University Press, Oxford
Boghossian P, Peacocke C (eds) (2000) New essays on the a priori. Oxford University Press, Oxford
BonJour L (1998) In Defense of Pure Reason. Cambridge University Press, Cambridge
Capellen H (2012) Philosophy without intuitions. Oxford University Press, Oxford
Casullo A (2003) A priori justification. Oxford University Press, Oxford
Cohen JA (1981) Can human irrationality be experimentally demonstrated? Behav Brain Sci 4(3):317–370
Devitt M (2005) The is no a priori. In: Steup M, Sosa E (eds) Contemporary debates in epistemology. Blackwell, pp 105–115
Frankfurt HG (1969) Alternate possibilities and moral responsibility. J Philos 66(23):829–839
Frappier M (2004) Heisenberg's notion of interpretation, dissertation available on Pro Quest
Gettier E (1963) Is justified true belief knowledge? Analysis 23:121–123
Harman G (1977) The status of morality. Oxford University Press, Oxford
Heisenberg (1983) The actual content of quantum theoretical kinematics and mechanics. NASA
Heisenberg (1927) Über den anschaulichen Inhalt der quantentheoretischen. Kinematik und Mechanik. Z. für Phys. 43(3–4):172–198
Klampfer F (2017) The false promise of thought-experimentation in moral and political philosophy. In Borstner B, Gartner S (eds), pp 328–348
Kuhn T (1977) A function for thought experiment. In: The essential tension selected studies in scientific tradition and change. The University of Chicago Press, Chicago
Lehrer K (1990) Theory of knowledge West View Press, Inc., Boulder, CO
Nolan D (2016) Method in analytic metaphysics. In: Cappelen H, Szabo Gendler T, Hawthorne J (eds) The oxford handbook of philosophical methodology. Oxford University Press, Oxford, pp 159–178
Peacock KA (2018) Happiest thoughts great thought experiments of modern physics. In: Stuart M, Fehige Y, Brown JR (eds) Routledge companion to thought experiments. Routledge, London, pp 211–241
Poincaré H (1958) The value of science. Dover, New York
Stuart M (2016) Taming theory with thought experiments: Understanding and scientific progress. Stud Hist Philos Sci 58:24–33
Stuart MT (2017) How thought experiments increase understanding. In: Stuart M, Fehige Y, Brown JR (eds) Routledge companion to thought experiments. Routledge, London, pp 526–545
Gendler TS (1998) Galileo and the indispensability of scientific thought experiment. Br J Philos Sci 49:397–424
Thompson JJ (1985) The trolley problem. Yale Law J. 1395–1415. http://library.du.edu
Weinberg JM (2007) How to challenge intuitions empirically without risking skepticism. Midwest Stud Philos 31(1):318–343
Williamson T (2013) How deep is the distinction between a priori and a posteriori knowledge? In: Casullo A, Thurow J (eds) The a priori in philosophy, Oxford University Press, Oxford, pp 291–312

Chapter 3
Understanding Thought Experiments

How should we understand TEs and the usual verdicts-intuitions resulting from them? In this chapter we briefly summarize the leading theories of TEs, hoping to do justice at least to the main ideas and arguments characterizing the present debate.

3.1 Empiricist-Inferentialist Proposals

Let us start with empiricist proposals, historically very prominent both in relation to scientific and more philosophical TEs.

The pioneering approach in the field is due to Ernst Mach. In his book *The Science of Mechanics*, first published in 1883, he discusses Stevin's chain TE, and concentrates upon a crucial question: why does Stevin feel that "the endless chain does not move" (Mach 1919: 26)? He describes his presumed feeling as a piece of instinctive cognition, and locates its origin in the process in which "/E/verything which we observe imprints itself uncomprehended and unanalysed in our percepts and ideas" (Mach 1919: 28). He is optimistic about the reliability of the imprinted material, assumes that it is "free from all subjectivity", and characterizes this accumulated experience as a "treasure-store" (Ibid.) from which our thought-experimenting might safely proceed.

The empiricist line has many followers on the contemporary scene. Kornblith (1998, 2008) and Devitt (2008, and elsewhere) both propose the view that TEs and the resulting intuitions represent empirical investigations and empirical judgments respectively. Both concentrate on human knowledge of natural kinds and take the empirical knowledge of empirical kinds (Devitt's mentions 'echidnas') as paradigm of the knowledge derived from TEs. Here is Devitt:

> …philosophers have acquired considerable knowledge of many kinds over a lifetime of acquaintance with them. The philosophers' intuitions that draw on this knowledge, draw on these theories of the world, are not a priori but empirical. The philosophers are conducting

N. Miscevic, *Thought Experiments*,
https://doi.org/10.1007/978-3-030-81082-5_3

thought experiments (…), counting themselves as experts about the kind in question. (Devitt 2008: 103)

For example, in epistemology one does a lot of thought-experimenting in order to determine what condition a given mental state should satisfy in order to count as a piece of knowledge. One typically pays attention to particular items, e.g. cases of accidentally true beliefs vs. non-accidentally true ones, and the like. Kornblith and Devitt assimilate this process to the gathering of samples, done, for instance, by a rock collector, "for the purpose of figuring out what it is that the samples have in common":

> When we appeal to our intuitions about knowledge, we make salient certain instances of the phenomenon that need to be accounted for, and that these are genuine instances of knowledge is simply obvious, at least if our examples are well chosen. What we are doing, as I see it, is much like the rock collector who gathers samples of some interesting kind of stone. (2008:10)

Devitt is particularly interested in the role of intuitions in linguistics, most importantly in probing the opinions of the native speaker (Is this a sentence of your language? To what does "he" refer to?). On his view of intuitions in general, "intuitive judgments are empirical theory-laden central-processor responses to phenomena, differing from many other such responses only in being fairly immediate and unreflective, based on little if any conscious reasoning" (2006:103;). Linguistic intuitions, in particular, "are opinions resulting from ordinary empirical investigation, theory-laden in the way all such opinions are (2006:98). "His first important move is to liken these cases to the cases of judgments about skills, e.g. judgments that a skilled typist can make about her own performance(s). His second important move is to stress that linguistic intuitions-judgments are just opinions.

Let me note a few typical difficulties for the general empiricist approach. First, to start with Mach's example, are we sure that the impossibility of perpetual motion is just the result of empirical imprinting. Could it not be a matter of a deeper innate scheme concerning movements of material bodies? Second, when we pass to more specialized intuitions like linguistic and mathematical ones, it looks like they come from competencies that are not due to empirical learning only. Devitt says explicitly that "(s)omeone who has the relevant competence has ready access to a great deal of data that are to be explained. She does not have to go out and look for data because her competence produces them" (2006: 106). And the whole motivation of Devitt's project was to get rid of immediate access to (deliverance of) competence. Now, Devitt replies that we "have immediate access to data but not to information about the data". I suppose he means at least the following: the judgment (or the pre-judgmental intuitional "feeling") to which the subject has first-person access might be part of the data her competence produces. But, first, she might not know that this is indeed a competence-produced datum, second, she has to re-interpret it anyway in order to formulate it in a judgment, and third, she cannot know how reliable the judgment is. There is a way out for the opponent. As to the first point, the subject might be default justified in taking her immediate feeling or judgment to be a valid datum, a record of what her linguistic skill would allow of prohibit. As to the second, the interpretation is minimal, involving the use of a minimal and pre-theoretical vocabulary. As to the third, she might be default justified in taking her immediate feeling or judgment to be reliable. Devitt's strictures are avoidable.

This points to a dis-analogy not between linguistic intuitions and pure perceptual judgments, but between linguistic intuitions and folk botanic and zoology. Devitt's assumes there is none, but there is. Echidnas, to use Devitt's favorite example, are not connected to one's mental apparatus to the degree one's language is. I am more familiar with dogs than with echidnas, but I don't get information about dogs just by rehearsing what I would do if…In contrast, my linguistic competence is not a separate entity, running around and having to be caught and studied. There is, moreover, little dis-analogy between linguistic and simple spatial perceptual judgments, crucial for many physical TEs. Our understanding of space might rely on competences that are largely innate, the way the linguistic ones are.

We now pass to the variant of empiricism concerning TEs that is most popular in philosophy of science. It is due to John Norton who has been arguing for it for decades in a series of brilliant papers (see Norton 2004, ……). TEs are mere arguments, he claims, only "disguised in a vivid pictorial or narrative form" (Norton 2004: 45), and the details of the form are irrelevant to the conclusion. Here is his somewhat caustic formulation. Thought experiments are arguments which:

(i) posit hypothetical or counterfactual states of affairs, and
(ii) invoke particulars irrelevant to the generality of the conclusion. (Norton 2004: 45).

He describes his theory as "an account of thought experiments as ordinary argumentation that is disguised in a vivid pictorial or narrative form. This account will allow me to show that empiricism has nothing to fear from thought experiments. They perform no epistemic magic. In so far as they tell us about the world, I shall urge that thought experiments draw upon what we already know of it, either explicitly or tacitly; they then transform that knowledge by disguised argumentation. They can do nothing more epistemically than can argumentation." (Ibid.)[1]

I find it incredible that most of the material typical for TEs is characterized by Norton as being "irrelevant" to its conclusion. But I don't want to be harsh. So, let me borrow for criticism of Norton some ideas from Sören Häggqvist who is

[1] Here is a longer quote: the thought experiments that interest me here are those of the natural sciences that do yield contingent knowledge of the natural world. According to empiricism, they can only do so if knowledge of the natural world is supplied to the thought experiment; that is, if this knowledge comprises a portion of the premises upon which the argument proceeds. It may enter as explicitly held knowledge of the world. We assert on the authority of an empirical theory, special relativity, that a moving rod shrinks in the direction of its motion. Or it may enter as tacit knowledge. We just know that the space of our experience never runs out; we have never seen a boundary in space beyond which we could not pass, unless there is already something past the boundary to obstruct us. I do not seek here to argue for empiricism; the debate between empiricism and other epistemologies is as ancient as philosophy itself and not likely to be advanced fundamentally here. However, empiricism is overwhelmingly the predominant epistemology in philosophy of science, so that an account that accommodates thought experiments to empiricism in a simple and straightforward manner ought to be accepted as the default, as opposed so some more extravagant account. I claim this default status for the view advocated here 2004:49.

more sympathetic to Norton's line than I am. For Häggqvist, the thought experiment plays a justificatory role in the same way a real experiment does: by contradicting or supporting a claim made by the theory through participation in an argument. This makes (thought) experiments parts of arguments. However, Häggqvist denies Norton's claim that the performance of a thought experiment just is the performance of an argument, because an experiment cannot be formally valid or invalid. Let me illustrate with Galileo's TE. I shall borrow the reconstruction from Häggqvist (1996), but I hasten to add that he is not generally an inferentialist; according to him there is more to TEs than just the argument. In connection with Galileo, however, He offers a simple looking argument:

(i) A cannonball and a musketball could be attached to each other.
(ii) If a cannonball were attached to a musketball, they would not fall both slower and faster than the cannonball alone.
(iii) If the speed of a falling body is proportional to its weight, then if a cannonball were attached to a musketball, they would fall both slower and faster than the cannonball alone.
(iv) The speed of a falling body is not proportional to its weight 1996: 114).

He then proceeds to give a formulaic expression to the argument, proposing that we let C stand for the (proposition describing the) situation imagined, T for the theory tested and W for the critical proposition falsified by C, and entailed by the theory in conjunction with C We than obtain the following general form;

$$\Diamond C$$

$$T \supset (C \,\Box\!\!\rightarrow W)$$

$$C \,\Box\!\!\rightarrow \neg W)$$

$$\overline{\qquad\qquad\qquad}$$

$$\neg T$$

In the inferentialist's view (not Häggqvist') this is all there is to the TEs. However, the obvious problem is with the first step, the possibility claim, in the first premise. It seems crucial to the whole argument, and it should be justified one way or another. The naive claim is that the experimenter just "sees" that is possible to tie the two balls together. And the regimented scheme depicts no argument *for* it, only argument *from* it. So, Brown is right in pointing out the particular(ized) character of the claim. It is a claim about a given, imagined pair of balls, a cannonball and a musketball. Now, the claim needed is not just that "a cannonball and a musketball could be attached to each other." It is rather that.

> (S) it is possible first, that they be attached to each other, second, that they will then fall together, and, third, that in falling together they will have one and the same and common speed.

This conjunctive possibility is crucial. (It would be very fine for a fanatical Aristotelian if musketballs routinely extricated themselves from the attachment in order to fall more slowly than their fat and heavy companions). We are prone to overlook it, since we find it so natural. But whence does the natural belief derive and what justifies it? I think Brown is right in stressing the manifestly non-inferential character of particularized modal claims and beliefs (note that his terminology is less technical). But the question remains. One line of answer would be to appeal to ordinary and implicit background knowledge, or, to be less committal, to body of firm belief. People believe firmly, albeit implicitly, that bodies attached to each other do fall, and indeed that they fall together, as a single body. They seem to accept that falling together, as a single body, implies having the same speed. Now, if there is such an implicit firm body of background belief available, the only two things immediately needed by the epistemologist are first, an account of some mechanism that mobilizes it, and second, an account of what justifies it (or makes one entitled to it), and possibly turns it into knowledge. (A third item needed less urgently might be some information about the origin of the body of background beliefs.

Given this situation, there is little hope in trying to reinterpret the discovery as a case of deductive learning. Even if we assume that there are purely propositional ways of recounting the reasoning performed in the experiment, which do not involve the use of imagining, or "picturing in front of one's inner eye", but consist of sentences or formulas only, this will not help. We need to imagine the propositional alternative as streamlined as possible, with reasoning made explicit regimented and formalizable, if not outright formalized. For one, I am not sure that such assumption makes sense at all, but the deductivist is committed to claiming that the reasoning in TE can be captured by such ways. The "deductivist" or "argumentationist" must claim that learning by TE in fact reduces to grasping dimly and maybe unconsciously something like the regimented alternative. This is clearly implausible. A list of propositions capturing all the assumptions needed might not be available for the inquirer who is inventing or performing a TE.(E.g. Stevin might not have known any other way to prove that the system he is describing is in equilibrium, than the way of his TE, but later day physicists have such other ways at their disposal). Furthermore, regimented alternative is heuristically of much less value than the "experimental" alternative. Second, the judgments immediately derived from (TEs) are modal and taxonomic particularized judgments. However, people are often better at assessing the modal and taxonomic features of a particular token situation, then at judging an abstractly described type. And lo and behold, our immediate, intuitional TE judgments are geared to singular, concrete situations. Finally and most importantly, arguments come after and not before (the imaginative parts of) TEs. The purely argumentationist escape is no solution.

3.2 A Platonist Alternative

3.2.1 *The Strong Platonist Option*

James Robert Brown

Let us now pass to the opposite extreme. Imaginative intuitions are windows upon Platonic heaven, James Robert Brown optimistically claims.

He thus follows and revives a very important tradition in understanding intuition. Historically, the first account of intuitional beliefs, famously offered by Plato, has been focused upon mathematical and (ostensive) moral truths and purported objects that this truths concern, i.e. Numbers and Forms. It further offered a tentative account of our sensitivity to the objects in the domain, which used the metaphor of seeing transferring the notion of perceptual capacity to intuitional one: the sensitivity is basically the capacity to "see" the domain with the eye of one's mind. The account is known as epistemic Platonism. Notice that it bypassed the issue of *our concepts*, i.e. of number or justice, and dealt directly with "seeing" objects and facts in the domain of reference. Brown is a staunch defender of such a line, in stark contrast with more recent and more fashionable approaches that place the apriority into our "concepts", to be briefly discussed in the next sub-section.

Consider now three classical puzzles concerning TEs, which Brown, I suppose, takes implicitly as supporting his optimism about "windows upon Platonic heaven". The first is abstractness: the contents of mathematical intuitions–considered in themselves independently of their physical applications–have seemed to many competent thinkers to refer to a kind of items situated outside the experiential, spatio-temporal universe, whose strangeness is conspicuous for mathematicians themselves. Numbers are supposed to be abstract, and numerical. properties as well. Counting apples in a bowl gives you their number. But apples-in-the-bowl form collections (or aggregates) i.e. concrete entities, whereas the mathematical orthodoxy claims that only sets, the abstract counterparts of collections, have numerosity (cardinality). Sets are abstract objects, and such objects are problematic for our approach, since they do not have instances.

The next is universality. Many intuition contents seem to be universally valid or true. The elementary arithmetic offers the best known example. As Frege kept pointing out, it is true about just every sort of object: you can count and add kings and cabbages, thoughts, properties, events and whatever you possible care to. Take next the so-called conceptual truths, for example, that things are never of two contrary colors, say red and green all over at the same time or that nothing has two lengths in the same respect (within the same frame of reference). Philosophers have generalized this fact of incompatibility. Call the generic property like color and length "determinable" and the particular instance, shade or particular length "determinate". Then it is a universal truth that determinates belonging to a common determinate exclude each other. Now, how can the finite human mind grasp and rationally believe such universal truths? How can we ever be sure of them if our only route to them is experience?

Finally, the intuition-contents of many intuitions seem to involve necessity. And, this apparently necessary character of intuition-contents is one of their central and most problematic features. Of course, necessity is just one from the family of modal properties, including possibility, impossibility, contingency and actuality. The standard philosophical view about the access to modal properties follows the common-sense points. First, if something–a state of affairs–is conceivable or imaginable it is taken to be possible in a very general sense. If it is inconceivable and/or unimaginable it is usually taken to be impossible, at least prima facie. If a negation of a proposition is inconceivable, then the proposition is prima facie necessary.

Passing to TEs themselves, let me summarize the main points of the Platonist argument (as presented in the section entitled "Why apriori?" of Brown 1986), regrouping them a bit so as to stress those elements which present the challenge to the empiricist.

The first point is that in the thought experiment "there has been no new observational data" (Brown 1986: 11). The second, armchair physics is "not a case of seeing old empirical data in new ways" (Ibid.) Also, we have the ease of the jump: "The burden of any thought experiment rests on the establishment (in the imagination) of a phenomenon. Once the phenomenon is established, the inference to a theory is fairly unproblematic; that is, the jump from data to theory is relatively small" (1986:4). In the next chapter, I shall try to show that a more naturalistic alternative, focusing on mental modelling in thought experimenting, can resolve the puzzles.

Brown is famous for having chosen and defended the Platonic option. It has two components, the epistemic and a metaphysical one, and it is the first that will concern us here. According to it, some of the successful TEs awake our intuition, which is a Platonic capacity that brings us in touch with abstract objects and truths about them. It yields a priori knowledge of such truths.

> In conclusion, my case for an a priori view of (some) thought experiments rests on three considerations. The first of these concerns facts such as that there is no new empirical evidence playing a role in thought-experimental reasoning. The second concerns the Platonic framework involving both mathematics and laws of nature, and how nicely thought experiments fit into this framework. (Brown 2004:1137).

Let me just mention for the sake of completeness that epistemological Platonism allows of a more imaginative i.e. quasi-perceptual approach, which likens the Platonic insight to perception and to visual imagining, and a more conceptual-propositional approach, which treats perception-like capacities as completely extraneous to the insight, which is a purely rational-intellectual matter. Similar contrast is to be found among anti-Platonists, both in more realist, and more anti-realist (Kantian) versions.

Brown's master argument for his view is finely summarized in his leading paper in the present issue. He stresses «the coherence of the three ingredients». The first of these is mathematical Platonism: mathematical entities are abstract and outside of space and time, and we can somehow intuit or perceive them. The second is the view that some TEs, such as Galileo's, give us a priori knowledge of nature. The third ingredient is the claim that laws of nature are relations among abstract entities. The first implies that we have a priori knowledge of some abstract entities. Now, if laws of nature are not mere empirical regularities, but rather relations among properties, that is, among abstract entities, then it is possible in principle to have a priori knowledge of abstract entities, and hence, of laws of nature. The intellectual grasping of a law of nature «happens when the mind's eye is focused in the right direction, which very plausibly happens in a thought experiment».

But can our alleged acquaintance with Platonic properties be explainBed? One challenge comes from the so-called Benacerraf's dilemma[2] and its more recent descendants. Take any arithmetical statement, for example, There are prime numbers greater than three", or even some simpler one, like "Two plus three equal five". If you take them literally, and accept that they are true, since they seem to be about numbers, it looks like presupposing realism about numbers: there are numbers and they have properties. However numbers are abstract objects, which cannot act upon our mind and cause our beliefs. So how can we know that p?

If you get desperate about the question, you can opt for a second-horn: anti-realism about numbers, and claim that numbers are our mental constructions, and that p refers to something mental. Then we have a lot of other problems, e.g. that it sound unnatural, that it is mysterious why mathematic is successful in natural science, etc.

Benacerraf talks about the problem of numbers causing our beliefs. But even without the notion of cause, one can reconstruct the same sort of problem. If numbers do not cause our beliefs, and we still have beliefs that seem to be about them, then they seem to be mind-dependent; how else could we have beliefs about them without them having causal influence? Here is one version of the dilemma:

(i) Platonism is committed to the acausality of mind-independence of mathematical objects.
(ii) Any causal explanation of our mathematical reliability is incompatible with the acausality of mathematical objects.
(iii) Any non-causal explanation of our mathematical reliability is incompatible with the mind-independence of mathematical objects.

Therefore,

(iv) There is no explanation of our mathematical reliability that is compatible with Platonism.[1]

Consider the options for a Platonist. She can take the original Plato's line and try to explain the (presumed) Platonic intuition, or he can refrain from the task. If he refrains, she can either take such intuition as a basic or primitive phenomenon, in particular, take intuition-capacity as a basic capacity, that itself explains a lot of things, but cannot be fully explained by our present means. This line remains within a general explanationist framework, but stresses the special, primitive place of intuition. Alternatively, she might give up on explanatory task, and rest content with a quietist resignation: there is nothing to explain. The contrast between the two no-explanation stances can be summarized simply: according to the first, there is

[2] Formulated in the famous paper by Benacerraf (1973).

something to explain, only we don't know how to do it, according to the second, there is nothing, and the very explanationist program is misplaced.

Brown seems to be a primitivist. But primitivism explains nothing, and if taken seriously, it comes dangerously close to anti-explanationism. It is plagued by.

Brown's primitivism postulating a primitive capacity. In his Proofs and Pictures he has clearly rejected the causal explanation. It would be interesting to know what he then thinks about the premise (iii), claiming that acausal explanations are bound to make mathematical objects mind dependent. Here is an additional point: if Brown thinks there cannot be any explanation at all, then he is more of a quietist anti-explanationist than a primitivist. And then it would be fine to know his reasons for giving up.

Let me then conclude this section with three problems for J.R. Brown. First, where do the initial, particular judgments in a TE (like "These two items tied together won't fall with the speed proportionate to their aggregate weight", or something similar) come from? They seem to precede the general insight, the "jump to theory", and not to derive from it. So, given that he takes the general insight to be Platonic, what is the nature of the preceding particular judgment? Is it Platonic as well? If yes, what is the story: does the thought-experimenter implicitly grasp universals in a platonic fashion and then apply the grasp to the particular situation at hand, to this cannon ball and this musket ball she is imagining? But then the jump to the general truth is implicitly performed before and not after the particular judgment. Or is she capable of noticing the truth of the particular judgment before arriving at the Platonic general insight, and independently of it? If yes, is the Platonic insight really needed?[3]

3.2.2 *The Conceptualist Moderate Platonist Approach*

Problems of the kind just mentioned have prompters some philosopher, sympathetic to the spirit of Platonism, to build a much more moderate version. This Moderate Platonist approach does not speak of Platonic ideas, but rather of concepts: the insights offered by TEs are insights into the nature of concepts playing the main role in them.[4] The main authors here are Jackson (1998), Chalmers (1996) and Peacocke (1998, 1999).

Let us start with an example. Take the Ship of Theseus TE. Here, we imagine the narrative of ship A with parts being replaced, the resulting ship B, still traveling, and

---

[3] Let me mention an additional problem about history of famous TEs. As far as Galileo goes there are serious problems with Platonism about his TEs. The research done by Stillman Drake (above all his (1978) book) has changed our picture of Galileo, and it now looks different - with stress on real experiment, exact measurement etc. Drake has shown that the material from *Dialogues* and *Discourses* is just the tip of the iceberg. Galileo has not revised his opinions just on the strength of few TEs, but thanks to the impressive amount of his actual, carefully performed experiments. (Generally, to make Galileo into a champion of platonic or aprioristic physics is now a bit implausible.).

[4] See for instance Boghossian and Peacocke (Eds) (2000), Szabo Gendler, T. and Hawthorne, J. (Eds.): *Conceivability and Possibility*, Oxford University Press, 2002.

the emergence of a ship C built on discarded original parts of A. Is A identical to B or C, we are asked. Next, we opt for one of them, say B, expressing the intuition that B is identical to A. Most people need not another analogous story, say of a knife, to conclude that the morals are generalizable. They somehow arrive at the general conclusion that continuity makes for identity through time. Similarly, in the Twin Earth thought experiment one first judges that XWZ is not water, and then generalizes to the view that composition is essential for stuffs.

Now, the Moderate Platonist proposes that the procedure concerns the concept or concepts of the relevant property or object in our examples, concepts MATERIAL OBJECT, PART and IDENTITY (we mark concepts by using capital letters). Concepts stand for senses, as Frege claimed: the concept MATERIAL OBJECT stands for the sense of the words "material object"; senses are not mental of psychological, but have a life of their own, and this makes conceptualist standpoint slightly Platonic.

The Moderate Platonist conceptualist also accepts a very important further assumption: whatever the nature of the concept of something, say MATERIAL OBJECT, the propositions that analyze it and offer a meaning of "P" are analytic and a priori knowable. The general conceptualist Moderate Platonist line for showing that conceptual knowledge is a priori and substantial relies on Frege's point according to which sense determines reference, and can be summarized as follows:

1. our knowledge of concepts is obtainable from the armchair, independently of experience.
2. concepts known are senses (definition of concept)
3. senses determine reference (Frege's point)
4. concepts determine (to a considerable extent) their extensions, therefore
5. our conceptual knowledge teaches us about extensions.

Less technically, our conceptual knowledge teaches us about reality, say the nature of ship as a material object.

Jackson and Chalmers have appealed to conceptual issues to defend their arguments against physicalism, in particular those relying on TEs. Let P be the conjunction of physical truths about the world, and let Q be a phenomenal truth like "someone is conscious" or "someone is having a such-and-such experience". But we can conceive of a physical being, physically like us, that lacks experiences and consciousness; a zombie-like creature. Having an experience cannot be deduced from physical facts, so, given our concepts, in particular the concept QUALE that stands for the qualities of experience, the zombie situation is conceivable. But then, physicalism might be false.

One of the most important items of Chalmers's agenda is to defend the apriorist view that our present intuitions about conceivability of particular scenarios, when the conceiving is done carefully and reflectively, can offer substantial insight into serious possibilities and impossibilities, and dictate in outline possible outcomes and rational lines of inquiry. He himself reminds us that "there is a long tradition in philosophy of using a priori methods to draw conclusions about what is possible and what is necessary, and often in turn to draw conclusions about matters of substantive metaphysics."

(Chalmers 2002:145), and places his research squarely within the tradition, Since he follows the tradition of tying conceivability very closely to concepts, the view amounts to accepting a strong determination of epistemic strategies by our present concepts (like for instance QUALE), which we have just mentioned. He then uses the framework obtained to argue from our present perplexities about qualia, some of which are tied to apparent conceptual impossibilities endemic to our best present QUALIA concept, that qualia cannot be in principle physicalistically accounted for, and that therefore they are in themselves non-physical.

Conceptualism has many problems. Tim Williamson has argued (in his 2007 book) that it misplaces the point of TEs: they are about the world, not about our concepts. Another problem concerns the nature of concepts. If one remains faithful to Frege's heritage, one ends up in serious Platonism, with the problems that go with it; if one wants to make the approach less Platonistic, one ends up in psychologism. And, if concepts are just psychological entities, they cannot teach us about the world in an a priori fashion, as the conceptualist would want them.

3.3 A Kantian Proposal

Is there a more plausible middle way between Brownian Platonism and Norton's infrentialist empiricism? Buzzoni (2008, 2018) claims there is, and offers the Kantian tradition as the guarantee of it. After Platonist, empiricist and naturalist-cognitivist proposals, he promises to come up with a Kantian view of both actual experiments and of TEs, in which both categories are brought close together within a slightly apriorist framework, but, fortunately, still kept distinct. The Kantian inspiration shows already at the outset: Buzzoni talks about "indeterminate multiplicity of categories individually constructed by the mind and tested operationally to solve specific problems" (15). Still, already here one wonders how much of (Kantian) apriority is preserved in the picture, in particular since the categories are empirically tested; some weak "relative apriority" might survive, but this seems hardly sufficient for an "reflexive transcendental conception" of scientific enterprise. The operational side is more explicit: we can empirically ascertain the properties of physical bodies only insofar we connect them to concepts and to reproducible actions that make the object interact with other objects. "These actions are always guided by concepts and values, and it is only for this reason that they have cognitive significance" (2008:23).

Buzzoni concludes his (2008) book by seven concluding take-home messages, that summarize the conception and intend to show its power. I will accompany the summary with my questions and critical remarks. Consider the first six points from the list.

First, according to Buzzoni, a TE is the anticipation "of a hypothetical experimental situation". Nothing to object, but not particularly challenging or new. However, if "anticipation" is taken literally, meaning that the though-experimenter really foresees that the corresponding real experiment will be performed, then it becomes quite problematic. Buzzoni gives some hints in direction of this problematic reading, by claiming that "practical-technical feasibility is thus the decisive criterion even for empirical thought experiments" (2008:108).

Second, experiments are connected to interpretation: "every new experiment (…) compels us to reinterpret from its point of view the whole history of that theory's experimental tests" (2008:129). Experiments and TEs "are born and develop within the history of their interpretation" (Ibid.). I must admit that "every new experiment" and "whole theory" in the penultimate sentence seem to me as an exquisite exaggeration. Does really every new experiment concerning a given theory force us to reinterpret the whole history of the theory? Do we rewrite every year the histories of well-known theories? Alternatively, Buzzoni might want us to count as "new" only revolutionary innovations. But then, his claim comes close to triviality: revolutionary innovations are precisely those that change our interpretations of the theory.

Third, both experiments and TEs have a "technical-operational nature" (2008:130); one can always connect TEs with experience. Well, if "connect" is taken in a very general sense, there is no problem; but the same holds for stories in holy books of world religion; one can always connect them with some experience in some way or another. In chapter Three a more precise formulation is offered: the causal sequence depicted by a TE , guided by the actual laws of nature, would be realized if the TE were. Well, it is not true that the composite musket ball-cum-cannon ball body would fall at the same time more quickly and more slowly than the cannon ball alone; so, some hedging would be needed at this point.

Fourth, "the technical device built by the scientist (…) are just what nature would be if there existed in it only the variables contained in the technical device" (Ibid.) The formulation is not the happiest one, but let us accept the point.

Fifth, TEs explore the empirical meaning of the theory tested. Perfect, if we agree what empirical meaning of a theory is.

Sixth, idealization and counterfactual nature is the necessary assumption of increasing precision in the real technical testing device. Moreover, this is just the purpose of TEs (Ibid.); the link with point one is clear. Here, I admit of being a bit baffled. Take Einstein's train whose speed comes close to the speed of light. Is the point of this counterfactual assumption really the wish to help build very quick devices competing with light in running very, very fast? Many TEs are not technically implementable; does it entail that they have no purpose? Now, assume we take the six morals with some grain of salt, and agree with most of them. The picture they offer is the one of ancillarity: TEs can always be connected to experience, they explore empirical meaning of theories, and the purpose that guides them is improving experimental technology. The picture is extremely empiricist, an ultra-Nortonian picture of the kind Buzzoni rejects.

The surprise comes with the point seven: TEs "are irreducible to real ones because the activity of formulating hypotheses (…) and anticipations is, as such, irreducible to physical reality" (2008:132). The anticipation at work in TEs "simply expresses the mind's essential ability to distance itself critically from every reality" (Ibid.) We already mentioned it, in the context of the alleged transcendental status of asking questions. Indeed, this is the only point that seems to justify the term "transcendental" for Buzzoni's conception.

Unfortunately, the reason he offers is hard to decipher. If the purpose of TEs is the improvement of technology, then the distance from reality cannot be too big; in particular it cannot be a distance from every reality. Second, one wonders how this applies to the most famous scientific TEs. Take Galileo's proposal to imagine tying together a cannon ball and a musket ball and to reflect about the speed of the resulting body. What is the distance to physical reality he is taking? (He is neglecting the resistance of the air, but not intentionally; this particular distancing will come later, not in the TE itself.) What about Stevin's hanging chain? How is he there distancing himself critically from reality? And is not a point, if not the point of the scientific enterprise to come to know about reality, more than to distance oneself critically from it (as one distances oneself critically from bad practices, say, at one's university)?

An important consequence of this unity and distinction of real experiments (which Buzzoni shortens to "REs") and TEs is that neither would be what it is outside their mutual relationship. On the one hand, despite their conditioned empirical power, TEs are very important in science for the following reasons: (1) at the most fundamental (transcendental) level, without TEs there would not be REs (we would not know how to put our questions to nature); (2) at the operational-methodological level, TEs, to the extent that they are based on well-established scientific facts and laws of nature, even if they are not realized or we decide not to realize them, can "inductively extend our knowledge" (Buzzoni 2008, 96) and support, at least provisionally, fresh scientific claims. On the other hand, without already realized TEs there would be no reliable empirical laws on which new TEs can base their anticipated answers to new questions. To sum up, they are complementary in a typical Kantian sense: "(empirical) TEs without REs are empty; REs without TEs are blind" (Buzzoni 2013, 100). see also 2018: 334.

Here is his characterization of his own understanding of the issues:

On the basis of this concept of the a priori, I developed an account that mediates between Brown and Norton by considering TEs from two distinct but connected points of view: transcendental and operational (Buzzoni 2007, 2008, 2011). On the one hand, the account is transcendental in a very usual sense of the word: TEs are the condition of the possibility of REs because, without the a priori capacity of the mind to reason counterfactually, we could not devise any hypothesis and would be unable to plan the corresponding RE that should test it. This capacity underpins the distinction in principle – a properly transcendental distinction – between TEs and REs. This distinction "cannot be suppressed, since it is the same distinction between the hypothetical-reflexive domain of the mind (which can always contradict itself) and reality (which can always occur and develop in only one way)" (Buzzoni 2013, 99–100).

But while there is a distinction in principle between TE and RE on the transcendental level of the a priori capacity of the mind to reason counterfactually, the exclusively functional character of the a priori entails that, on the operational-methodological level, TEs—considering their empirical content and actual performance within a discipline—are and function essentially like REs (cf. Buzzoni 2013, 98).

This is the right point to raise a question about "reflexive transcendental" aspect of the conception proposed. The term "reflexive" is a useful reminder that doing science requires a lot of thinking; but the reader would miss a clearer picture of the divide between spontaneous, first order thinking and the reflection in TEs. One way to go would be to put the later at the level of generalization from some particular thought-experimental datum, although generalizing is often spontaneous, and has been taken to be such from Aristotle on. Another way would be to place it on the level of reflective equilibrium, integrating thought-experimental results with each other, and with the rest of science. But what makes the whole enterprise "transcendental"? Is it that TEs make explicit the conceptual framework that makes possible scientific testing; somewhat along the lines of M. Friedman's Kantian proposal? In Kant, "transcendental" has, of course, stronger meaning, having to do with possibilities of phenomena as such. Does Buzzoni mean this? At any rate he does not tell us much; perhaps a definition would help, as well as some dialogue with contemporary Kant-inspired work.

To conclude, the account is very useful to the reader interested in TEs; it offers a wealth of highly interesting and relevant historical material, skillfully placed in the context of contemporary debates. On the other hand, author's "operational and reflexive transcendental conception" is not made sufficiently clear; to make it more explicit is the task for the author, that might give us one day a new and original view of experimenting in thought.

3.4 Imaginative-Counterfactual Approach

Thought-experimenting involves proposing and considering counterfactual scenarios. What if we had a chain with such-such-characteristics? What if you managed to catch a beam of light? When considering the scenarios and responding to the leading question the experimenter normally uses her imagination. In few last decades a trail of investigations stressing the two traits, counterfactual object of and imaginative method in thought-experimenting has developed. In this section we shall concentrate on a recent proponent of the approach stressing this combination, Tim Williamson and in the next pass to a more concrete and detailed approach that also stresses the two features, but places them in a particular cognitive context, the one of mental modelling.[5] Williamson has written a lot about methodology of philosophy (the main work being his 2007 book), a big part of it is related to TEs. Here we have

[5] Let me just point to the work of Sorensen (1992), who has in early nineties already discussed the counterfactual nature of TE-scenarios.

to choose, for reasons of space, so we shall concentrate on his relatively recent and very innovative proposal about "knowing by imagining" (2016).[6]

He is careful to place the consideration of TE-counterfactuals in the context of ordinary reasoning about possible situations, actions and outcomes. In a recent paper he invites us to imagine a hunter who finds his way obstructed by a mountain stream rushing between the rocks:

> He reaches the only place where jumping over might be possible, and now considers what to do. If i try here, what is it going to be like?, he is asking himself. In order to answer the question, he has to consider alternative scenarios: what if he fails and gets hurt, or even drowns in the river? What if he embarks on a lenghty search of some alternative crossing spot? And he imagines various counterfactual scenarios. Tim Williamson 2016:117.

The cognitive qualities needed are certainly—rational responsiveness to evidence. as Williamson notes (2016: 116) and the capacity to develop adequate scenarios:

> the imagination develops the scenario in a reality-oriented way, by default.
>
> under suitable conditions, the method constitutes a reliable way of forming a true belief as to what would happen in hypothetical circumstances. ((116–117)

Interestingly, for Williamson fiction is not central for imagination. He prefers to stress the importance of practical matters:

> Even in the modern world, important decision-making often relies on knowledge or beliefs acquired through the imagination. For example, you are looking round a house, wondering whether to buy it. You want to know whether, if you lived in it, you would like doing so. You voluntarily imagine the antecedent of the conditional; your final decision may depend on what consequent you involuntarily imagine. 2016:117

On Williamson's cognitive view of imagination its main cognitive functions are raising possibilities and assessing the truth-values of propositions illustrated by Williamson through his hunter story 2016: 115).

The first general issue that arises here is: what is imagination and what is the role of image in it? How close is it to belief? The term "cognitive" seems to suggest a very high degree of closeness; what about the differences? Williamson's Hunter is creating a visual-kinesthetic moving picture, a video: "It will be like this", he tells himself.

I shall call such a video a mental model of the situation in the consequent, but this is not Williamsons way. I asked Williamson what is his full picture of imagination and how far would he go in downplaying the concreteness of (possibly imagined) examples? Can logic really take over the task? In discussion, Williamson kindly answered in the negative: logic is the means for describing the mental process "from the outside", not a picture of what is happening in the thinker's head. This keeps Williamson away from inferentialism of Norton's type. For him, imagination guards its specificity in relation to with normal fully abstract thinking Let me note my

[6] Thanks go to Tim for discussions at yearly conferences in Dubrovnik.

agreement with Williamson on his criticism of the inferential picture of judging about contrary-to-fact situations:

> The treatment of the process by which we reach counterfactual judgments as inferential is problematic in several ways.
>
> …the putative reasoner may lack general-purpose cognitive access to the auxiliary premises of the putative reasoning. In particular, the folk physics needed to derive the consequents of counterfactuals such as (6) from their antecedents may be stored in the form of some analogue mechanism, perhaps embodied in a connectionist network, which the subject cannot articulate in propositional form. (2016:145)

Here are the main steps in Williamson's argument against specificity of armchair judgments and knowledge:

1C. Many TEs essentially involve counterfactual reasoning.
2C. Counterfactual reasoning essentially depends on judgments of overall similarity of possible situations-worlds to the actual one, and on the capacity to judge interdependency of changes.
3C. The judgments in 2C.depend on general empirical knowledge.
4C. In typical TEs one cannot separate the contribution of general empirical knowledge from domain-specific (mathematical, moral, and probably epistemological normative etc.) contribution. Therefore,
5C. Typical products of TEs (particular and general armchair judgments) have a holistic origin, contrary to the traditional view of armchair knowledge. Their epistemic status is more empirical than the view would admit.

One weakness of the argument lies in 1C: the conclusion of a TE does not have to take the form of a counterfactual conditional. Given Williamson's admission that some cases considered in TEs might be real, an indicative conditional would do as well; but then the interaction of empirical sources and presumed rational ones might look different. However, Williamson has also a more ambitious goal in the book, namely show that modal knowledge reduces to counterfactual knowledge. It this is successful than it might give a stronger argument for C1. But for this he does need some very strong assumptions in order to derive the rules for necessity operator (in Appendix one), assumptions that do not seem to capture the ordinary, or even the philosophical understanding of what it is for a proposition to be necessary.[7]

Williamson writes that "much of our knowledge of the future depends on cognitive processes relevantly similar to imaginative exercises. 2016:120. Judgments are easy to elicit with concrete examples, and with naive subjects it is the only way. His main example are Gettier TEs whose main point is captured in the following counterfactual:

If there were Gettier cases, they would be cases of justified true belief without knowledge.[8]

But how does one figure out what facts not explicitly mentioned in the formulation of the TE should hold in the imagined world? "By using the counterfactual

[7] Thanks go to Edi Pavlović for discussing these matters with me.

[8] In symbols: $(3*)\exists \times \exists_p GC(x, p) \Box\!\!\rightarrow \forall_x \forall_p (GC(x, p) \rightarrow (JTB(x, p) \& \neg K(x, p)))(185)$.

conditional, we in effect leave the world to fill in the details of the story, rather than trying to do it all ourselves", says Williamson. So, we imagine a Gettier situation as similar to situations in our surroundings, and then ask whether the hero of the story has knowledge or not. It is here that holistic considerations enter the story:

We can schematize the process of evaluating a counterfactual conditional thus: one imaginatively supposes the antecedent and develops the supposition, adding further judgments within the supposition by reasoning, off-line predictive mechanisms and other off-line judgments. All of one's background knowledge and beliefs is available from within the scope of the supposition as a description of one's actual circumstances for the purposes of comparison with the counterfactual circumstances (in this respect the development differs from that of the antecedent of an indicative conditional). Some but not all of one's background knowledge and beliefs is also available within the scope of the supposition as a description of the counterfactual circumstances, according to complex criteria (the problem of co-tenability). To a first approximation: one asserts the counterfactual conditional if and only if the development eventually leads one to add the consequent (143).

The main line of counterfactual approach seems acceptable, and I find Williamsons interest for the working of imagination very congenial. But, what is the mechanism that enables us, humans, to go into detailed, counterfactual representing, and what competences play the crucial role here. In order to answer this question, I think that we need to supplement the counterfactual considerations with a more detailed, cognitive picture. To this alternative we now turn, with apologies for brevity.

3.5 Mental Modelling Approaches

The mental model alternative starts from the idea that in a typical TE, the thought-experimenter is encouraged freely to imagine things and then rearrange them. The view is then that these encouraged activities are distinctive of and crucial for TE. The view that TEs are such sophisticated "re-modellings in the head" is closed to the imaginative-counterfactual approach, since mental models are normally related to imaginative thinking, and in particular to imagining counterfactual situations. The approach has been proposed simultaneously and independently by Nancy Nersessian (1992; but see also her other works in References) and the present author (1992). It belongs to the non-Platonist tradition since it takes TEs to be processes of reorganizing or rearranging the old pieces of knowledge—old experiences, beliefs and the like. (The Platonist sees TEs as cases of simple seeing with a "mental eye". This overlooks a significant amount of imaginative operations involved. I concur with the tradition in believing that the insight yielded by some of TEs is to a large extent a posteriori. Namely, the non-mathematical material with which she works is experiential, and so her conclusions will be, in the last resort derived from experience. Her physics will be of the usual, a posteriori sort. The possibly a priori part is a priori in the sense of being derived from innate sources, which are themselves to be accounted for in terms of causal processes that have lead to them. In my first paper on the topic

(1992), I proposed that performing TEs consists in a particular way of manipulating a certain kind of representations. The representations and operations on them have a quasi-spatial quality, which distinguishes them from their counterparts couched in purely verbal or propositional medium. Although the basic mechanism enabling the construction and manipulation of the representations might be computational, the most important feature of representations and operations is precisely their concrete and quasi-spatial character. This has earned them the name of imaginative "mental models". Mental models are typically involved in understanding stories, or in ordinary planning of activities. When a reader encounters a description of a situation, she builds a model, a quasi-spatial "picture" of it. As new details are supplied by the story-teller, the model gets updated. The background conditions are dictated by the thought experimenter's general knowledge about the world.

A number of researchers have investigated reasoning in mental models. Johnson-Laird noted that "people may mentally represent possibilities by constructing "mental models" (1983); I actually got enthusiastic for the mental modelling approach by listening to his lectures in Trieste some decades ago. Other interesting works are Byrne (2005) and Stenning and Lambalgen (2008). The latter defines model in the following useful way:

> Roughly speaking, a scale model of X is an object which is structurally similar to X but suitably smaller. "Similar" and "structurally" have their usual context dependence as much as does "smaller"—in any particular case, the goal implicit in "suitable" will determine the contextual parameters for each. (2008:51),

He notes two important assumptions that are in force when conclusions about the target are drawn from characteristics of a scale model. The first is "that structurally similar objects will display the same behavior in structurally similar circumstances." And the second is that if some conditions follow from the ideal case, they will follow *mutatis mutandis* from an approximation to the real case.

How do mental models work? We shall talk more about the question in the next chapter. Here, just for warming up, let me offer an example, Putnam's Twin Earth as a source of application, since it contains exactly what we need:

> For the purpose of the following science-fiction examples, we shall suppose that somewhere in the galaxy there is a planet we shall call 'Twin Earth'. Twin Earth is much like Earth; in fact, people on Twin Earth even speak English. In fact, apart from the differences we shall specify in our science-fiction examples, the reader may suppose that Twin Earth is exactly like Earth. Twin Earth is very much like Earth; in fact, people on Twin Earth even speak English. In fact, apart from the differences we shall specify in our science-fiction examples, the reader may suppose that Twin Earth is exactly like Earth.
>
> He may even suppose that he has a Doppelgänger - an identical copy – on Twin Earth, if he wishes, although my stories will not depend on this. (Putnam 1975:223).

How do we imagine the scenario? The mental model account claims that we have in our memories a lot of material that is typical for our earthly surrounding. We use this material to construct a provisory, sketchy mental model of Twin Earth, a place very much like ours. I would imagine a seaside, similar to the Rijeka spot in Croatia where I live, with a mountain across, visible from my window. In my memory, I assume, stretches of see are labeled as "water", salty for that matter.

Now, in the story comes the point of difference with Earth. There are some linguistic differences, Putnam says, and they themselves depend on some of the peculiarities of Twin Earth. Next comes the crucial point:

> One of the peculiarities of Twin Earth is that the liquid called 'water' is not H_2O but a different liquid whose chemical formula is very long and complicated. I shall abbreviate this chemical formula simply as XYZ. I shall suppose that XYZ is indistinguishable from water at normal temperatures and pressures. In particular, it tastes like water and it quenches thirst like water. Also, I shall suppose that the oceans and lakes and seas of Twin Earth contain XYZ and not water, that it rains XYZ on Twin Earth and not water, etc.(Ibid.)

The crucial point is that the liquid called "water" is "a different liquid" from ours, the composition of which is indicated by a shorthand, XWZ. So, what do I do? In my mental model, with a nice stretch of sea borrowed from my everyday experience, I go for a seemingly small, but revolutionary change: the stretch of see is not labeled as "water" any more, it gets the label of the different liquid, XYZ! In fact, all the water around, in rivers rains and the like, will get re-labeled. I can now in my mental model represent the next step taken by Putnam:

> If a spaceship from Earth ever visits Twin Earth, then the supposition at first will be that 'water' has the same meaning on Earth and on Twin Earth. This supposition will be corrected when it is discovered that 'water' on Twin Earth is XYZ, and the Earthian spaceship will report somewhat as follows:
> 'On Twin Earth the word "water" means XYZ:

Indeed, now that I have re-labeled the stretch of sea, and the rest of water, I can imagine the spaceship, I can re-enact the reasoning of its crew members, and understand how they do the same I did: they first unthinkingly keep the label "water" for the liquid they encounter; once they realize (discover, come to know from chemists on Twin Earth), that it is a different stuff, they do what I did. Now, the stuff is relabeled as XYZ, and I can empathize with the report of the Earthian spaceship. Indeed, the word "water" here does not mean water; it means XYZ.

The same for other variations of the TE: temporal with water through centuries, and then molybdenum/aluminum switch.

We can now see that the crucial moves within mental model have been dictated by Putnam's formulation of the TE: the liquid on Twin Earth is '*not the same liquid*' as here. Imagine that the formulation has been different:

> One of the peculiarities of Twin Earth is that the water on the planet is not H_2O; its chemical formula is very long and complicated. I shall abbreviate this chemical formula of their water simply as XYZ.

Some people, lovers of chemistry, would of course react: If it's not H_2O, then it's not water! But many people would take it just as a part of the story (like we all take the idea in the Swan lake that swans transform into princesses and the other way around. Imagine an ordinary subject who has swallowed the different formulation. What she does not with her mental model might be that she leaves it as it is, if she doesn't care for chemistry at all. If she is at least a bit meticulous, she will add the label XYZ to the existing label "water". Now, if you ask her Is the Twin Earth water water? She might reply Yes, of course. Nothing essential has changed!

We can be more charitable to Putnam, and propose something weaker: it is a default assumption in people's mental models that the stuff is determined by composition, and that if you change composition you change the stuff (the same with swans: if a swan 'becomes' a princess it is not the same animal any more, in fact it has been replaced by a human being). But the default can be overridden: if the narrator insists, the listener will opt for non-default action of leaving the label "water" for the Twin Earth stuff.

What about the modal component? Well, our story enables us to find a niche both for essentialist intuition (our default assumption), and for our counterfactual intuitions, at least partially: if Twin Earth world were the actual world and dictated what is water, then Twin Earth liquid would be water (what about our water then? A difficult question).

One can see that a lot depends on formulation of the TE scenario, and that details in formulation could result into upheavals in the mental models constructed. (we shall say more in Chap. 5).

So much as a foretaste of the mental modelling alternative. Since I believe it is the most fruitful, and both scientifically and philosophically acceptable option, I will develop it in much more detail in the next chapter.

3.6 Conclusion

There is a wealth of proposals for understanding TEs. Here, we have distinguished on the one hand epistemological Platonism and epistemological anti-Platonism. The first has more conceptual and more quasi-perceptual varieties, and we have concentrated on the later, with J.R. Brown as the main theoretician. On the anti-Platonist side we have distinguish strongly realist option from more anti-realist, Kantian side. Here is the whole classification.

	Epistemological platonism	Epistemological anti-platonism	
		Realist	Anti-realist
Imaginative, quasi-perceptual	J. R. Brown	Hilbert, Mach, Williamson, mental modelers (including the present author)	Kant and the Kantians
Intellectual	Frege and neo-Fregeans	Norton, Devitt	conventionalists

We have sketched our short criticisms of some of the approaches and will not repeat them hear; we prefer letting reader decide for oneself. On the more intellectualist side, we have stressed Norton's proposal, and among the imaginative realists we talked about Tim Williamson's counterfactual approach and started the presentation of mental modelling theories. To these last theories we now turn.

Note

1. The version has been proposed by John Divers and Alexander Miller (1999); it is quoted here in the formulation given by Sosa (2002: 370).

References

Brown JR (1986) Thought experiments since the scientific revolution. Int Stud Philos Sci 1(1):119–128

Benacerraf P (1973) Mathematical truth. J Philos 70(19):661–679

Brown JR (2004) Peeking into Plato's Heaven. Philos Sci 71(December):1126–1138

Boghossian P, Peacocke C (eds) (2000) New Essays on the A Priori. Oxford University Press, Oxford

Buzzoni M (2007) Zum Verhältnis zwischen Experiment und Gedankenexperiment in den Naturwissenschaften. J Gen Philos Sci 38:219–237

Buzzoni M (2008) Thought experiments in natural science, an operational and reflexive transcendental conception, Königshausen & Neumann

Buzzoni M (2011) On Mathematical Thought Experiments Epistemologia XXXIV, pp 61–88

Buzzoni M (2013) On thought experiments and the Kantian a priori in the natural sciences: a reply to Yiftach J.H. Fehige, Epistemologia

Buzzoni M (2018) Kantian accounts of thought experiments. In: Stuart M, Fehige Y, Brown JR (eds) Routledge companion to thought experiments. Routledge, London, pp 327–340

Byrne RMJ (2005) The rational imagination how people create alternatives to reality. Bradford Book, The MIT Press, Mass, Cambridge

Chalmers DJ (1996) The conscious mind. In Search of a Fundamental Theory. Oxford University Press, Oxford

Chalmers DJ (2002) Does conceivability entail possibility? In: Gendler TS, Hawthorne J (eds) Conceivability and Possibility. Oxford University Press, Oxford, pp 145–200

Devitt M (2008) Ignorance of language. Oxford University Press, Oxford

Drake S (1978) Galileo at work. University of Chicago Press, Chicago

Divers J, Miller A (1999) Arithmetical Platonism: reliability and judgement-dependence. Philos Stud 95(3):277–310

Häggquist S (1996) Thought experiments in philosophy. Stockholm: Almqvist & Wiksell
Jackson F (1998) From Metaphysics to ethics: a defence of conceptual analysis. Clarendon Press, Oxford
Kornblith H (1998) The role of Intuition in philosophical inquiry: an account with no unnatural ingredients. In DePaul and Ramsey, 129–41
Kornblith H (2008) Knowledge and its place in nature. Clarendon Press, Oxford
Mach E (1919) The Science of Mechanics, trans. Thomas J McCormack, La Salle, IL (1919)
Nersessian N (1992) J. "In the theoretician's laboratory," PSA: proceedings of the Biennial meeting of the philosophy of science association 2:291–301, East Lansing: Philosophy of Science Association
Norton JD (2004) Why thought experiments do not transcend empiricism. In Hitchcock C (ed), Contemporary debates in the philosophy of science, Blackwell
Peacocke C (1998) Implicit conceptions, understanding and rationality. Philos Issues 9:43–88
Peacocke Being known. Oxford University Press, Oxford (1999)
Putnam H (1975) The meaning of "Meaning" in mind, language and reality, pp 215–270. Cambridge University Press
Stenning K, Lambalgen MV (2008) Human reasoning and cognitive science, a bradford book. The MIT Press , Cambridge, Massachusetts
Sorensen RA (1992) Thought experiments. Oxford University Press, Oxford
Sosa E (2002) Reliability and the a priori. In Szabo Gendler T, Hawthorne J (eds) Conceivability and possibility. Oxford University Press, Oxford
Van Fraassen B (2008) Scientific representation. Oxford University Press, Oxford
Williamson T (2007a) Philosophical knowledge and knowledge of counterfactuals. Grazer Philos Studien 74:89–123
Williamson T (2007b) The philosophy of philosophy. Blackwell Publishing, Oxford

Chapter 4
A Theory of TEs

4.1 Mental Modelling

We have briefly considered several contemporary approaches to TEs, with two extreme views, the inferentialist-empiricist and the strong Platonist one, and with several options in between. We now pass to the theory we find most plausible, the mental modelling theory.

The theory suggests that performing TEs consists in a particular way of manipulating a certain kind of representations. They might involve language-like representations and inference and computation on them, but typically, they involve more concrete representations, such as are used in imaginative operations. For example, some of imaginative representations and operations on them would have a quasi-spatial quality, which distinguishes them from their counterparts couched in purely verbal or propositional medium. We are invited to imagine a concrete situation, for instance being in an elevator in a free fall, or in the one that is accelerated upwards, and ask what we would feel.

Although the basic mechanism enabling the construction and manipulation of the representations might be computational, the most important feature of representations and operations is precisely their concrete and quasi-spatial character. This has earned them the name of imaginative "mental models", and many cognitive scientist argue that they are present overall in our mental life. We have mentioned that the classical work on mental models is Johnson-Laird, P.N. (1983).

Very similar to building such mental models is modelling through simulation, when one is invited to imagine oneself in a given situation. For example, confronting the Trolley problem TE, I might ask myself: what I were standing at the intersection of two railway roads, being able to manipulate the switch determining the road that a train will take, and if I were confronted with a dramatic situation in which I can save five people by sacrificing two? And I might imaginatively "simulate" the decision making. We shall return to simulation later, after we first describe the modelling done by building of mental models.

N. Miscevic, *Thought Experiments*,
https://doi.org/10.1007/978-3-030-81082-5_4

In the building scenario, when a reader encounters a description of a situation, she builds a model, a quasi-spatial "picture" of it. As new details are supplied by the story-teller, the model gets updated. The background conditions are dictated by the thought experimenter's general knowledge about the world.

Given the notion of the model we can place both the physical insights and the geometrical learning into a wider context of ordinary modeling and of spatial imagination ("intuition"), thus revealing the continuity between them: if ordinary beliefs about spatial relations are evidentially grounded in visualizing, and geometrical beliefs are sufficiently analogous to ordinary spatial beliefs, then geometrical beliefs might be also evidentially grounded in visualizing. There is an additional bonus in the offing: ordinary spatial beliefs have been extensively discussed both in the philosophical literature and in the psychological one, so that there are candidate cognitive psychological explanations of them; then, if the analogy between ordinary spatial and more specifically geometrical insights holds, one could use the same kind of explanation for geometrical beliefs, and integrate them into the general picture constructed by the joint efforts of philosophers and psychologists. According to it, the use of mental models demands rules for manipulation. Some constraints on manipulation come directly from the geometry of the model. It is essential that manipulation of elements mobilizes the spatial skills of the subject, his "knowledge how" which is generally not verbalisable. So, the first task undertaken is recapitulating the phenomenology of visual thought experiments, and distinguishing stages within a typical experiment in which some geometrical proposition is being imaginatively verified. Roughly, in the first stage the cognizer tries to imagine a scene verifying a given proposition. Sometimes a second, testing stage follows in which one tries to imagine situations that would falsify some given proposition. In the last, recapitulating stage, the cognizer typically first judges that no imagined situation falsifies the proposition, second, assumes that she has inspected all imaginable situation, and third, infers that it is impossible that the proposition tested is false, i.e. that it is necessarily true.

It is in the first phase that the cognitive theorizing deploys its explanatory force. It has shown that imagination and perception share so many traits that the use of imagination can be understood as the continuation of perception by other, rather similar, means. Imaginative capacities are to be explained in terms of their perceptual evolutionary origin; the upshot is that the knowledge of possibility is not opposed to the knowledge of actuality but derives from evolutionary acquired knowledge about actual spatial organization. Human imaginative capacities seem to work with elements from actuality, recombining them freely to yield (models of) possible situations In other words, although possible states of affairs do not act upon us, we are able to imagine them by exploiting the resources that did develop under the causal impact of the actual world, and of its concrete denizens. This further suggests that, on the normative side, the naive cognizer is prima facie justified in trusting her imaginings, to a large extent because her capacity had been subjected to empirical testing in most anodyne tasks and has withstood the tests. Finally, the way necessity is introduced in geometrical reasoning shall be briefly discussed. The view proposed is that there is no mystery involved, just a generalization over the imaginary situations "inspected", which is less reliable than the particular imaginings since it relies on the questionable assumption that all possible scenarios have been imagined. In short, the Platonists have discovered an important cognitive capacity, but have misinterpreted its nature, origin and rationale. The quasi-visual intuitions featuring in the traditional apriorist discourse can be to a large extent explained on naturalistic grounds.

Mental models can do most or even all of the explanatory work required. Remember Brown's proposed explananda we listed in Chap. 3: learning new things without new observational data, superiority of TEs to deduction in terms of heuristic power, obviousness and ease, and the ease of the jump anti-inductivist considerations. Add to these precision of geometrical intuitions and universality and necessity, characteristic of conclusions of many TEs.

Let us start with purely spatial or geometrical intuitions, since they are such a clear case, and are as well of fundamental importance for physical TEs; the obvious Brownian example being Stevin's TE. But let me revert to the simplest and the most famous case. At one point in the Platonic dialog, Meno's slave is confronted with the fact that by folding a square over one of its diagonals, the opposite sides will come to coincide. We can change the historical context a little bit, and replace actual cutting of the sloppy drawing of a square in the sand, by his looking at the drawing and imagining the square being folded over one of its diagonals. "How many figure will you get", Socrates asked. "Two", answers the slave. "Which one is bigger?" "None, they are exactly equal", replies the slave. In the first phase of such a visual exercise–the one crucial for us–the cognizer, here Meno's slave, uses his imagination to vary the relevant parameters. In the geometrical example it is the shape of the square. He either samples one or several imagined configurations, or lets an "inner movie", a dynamical model, unfold in his imagination. The imagined configuration is part of what we here call "mental model". We can follow Brown and see in these geometrical exercises a kind of geometrical TEs.

Now to the Brownian challenges. Remember that his first challenge was introduced by the claim that no new data are supplied by TEs. The features of the use of mental models that are most important for understanding how this use answer the challenge are probably the following: first, many problems, particularly those involving spatial relations, are much easier to solve in mental models, then verbally. Second, manipulating mental models gives the thinker a "view" of changes of situation. It is close to experience, not only in respect of its "feel", but, much more importantly, in respect of its concreteness, specificity and spatial character. Finally, very often, the solution of the problem is "imaged" (or "seen"), and this is just due to the peculiarity of the medium.

To illustrate the last point briefly, consider visualizing in geometry that can be taken as a limit case of building a model in one's head. Assume Meno's slave visualized a square (an in this sense built a geometrical "mental model" of it) and imaginatively folded it over its main diagonal. The visualization has enriched the slave's data base by making him aware of pieces of knowledge which he had no idea of deploying before, and items of practical know-how which can be manifested only in concrete manipulation. So, against the claim that no new data are introduced through TEs it should be pointed out that the mobilization of the previously unavailable knowledge does constitute an important widening of data base and a vital part of the process of discovery. Now the TEs from Galileo to Einstein typically bring geometrical intuitions to bear upon issues in physics. They help the inquirer and the reader mobilize the vast resources of practical spatial-geometrical knowledge—the data which are old in the sense of being "implicit" in the resources, but new (or "feeling" new) in the sense of not having been at inquirers' conscious disposal, or not even being in principle verbalisable for them.

Let us therefore consider the geometrical case in a bit more detail. The point of the exercise of visual variation is to get acquainted with various spatial-geometrical possibilities and alternatives. It is here that philosophers rightly become curious and inquisitive. Is the naive cognizer entitled to trust her imagination and why? "Are you sure that the two figures are equal?", we imagine Socrates asking. "Couldn't they sometimes be unequal?".

"No, that's impossible!", answers the slave with self-assurance atypical for his class. What happened?

A typical naive cognizer answers such questions by appealing to certainty and impossibility to imagine things otherwise. If pressed further, she will often say that she can "see" that the truth in question holds necessarily. A more experienced cognizer, or a theoretician, will appeal to the reliability of intuitions. These are the data that the normative discussion should take into account. The obvious answer is that she can know from experience that she is reliable about imagined simple shapes; this capacity has been tested throughout her life, and has passed the exams with flying colors. To generalize the point, if the use of spatial reasoning is as ubiquitous and successful as we have seen, we may conjecture that the deliverances of spatial imagination/ intuition are being routinely tested in everyday life or science. The practical success on the tests creates a presumption in favor of serious imagination and the attendant intuitions, that extends to all sorts of spatial knowledge. Modal assumptions about physical or spatial possibilities are often tested by actual findings and those that pass the test are retained. The policy of trusting spatial intuitions in their pure, geometrical guise and in their modal deliverances is then not as naive as some philosophers would like us to believe. The ordinary cognizer is internally justified in prima facie relying on her imagination in thought experiments, since her imaginative capacities have been tested and vindicated empirically, in the normal course of life in which controlled imagination is used in practical matters. This justifies internally the trust in particular episodes of imagining. Also, if the hypothesis proposed here holds, then there is an external, reliabilist justification available for geometrical intuitions that has a perceptual grounding. The epistemic power and success of imaginative intuitions are then due to the link with perception: imagination follows spatial perceptual constraints and is therefore a good guide to actuality. Finally, the very compellingness of intuition is a prima facie ground for trust: if we cannot but trust them, we are free to do it (ought implies cannot, therefore, "cannot mistrusts" implies "may trust").

In order to pass from geometry to physics, let me just mention a mixed, geometrico-physical example. You implicitly know that a circular chain retains its topological properties while sliding. But unless you are a trained scientist or philosopher of science, you have no way to express this knowledge in simple propositional way. Not even Stevin could have done it easily. So, he and his public do not have an adequate propositional representation at their disposal, and the implicit knowledge they have stays dormant and idle. However, with quasi-spatial model the implicit knowledge becomes usable. Let us now return to the central physical TE in Browns account, i.e. the Gallileo's one concerning free fall. The first part of Galileo's experiment concerning free fall involves reasoning about the role of weight and culminates in the intermediate conclusion that weight plays no role in determining speed. Simplicio is asked first to consider a light body L and a heavy body H falling freely, and then to consider a package P, made of L and H tied together. He is supposed to "become aware" of the contradiction present in the initial assumption that speed is proportional to weight. How does he achieve this? How are the presumed "new (quasi-)experiences", like vividly imagining the musket ball and the cannon ball falling together, constructed in her mind?

In terms of mental models this is how he does it: she first builds two smaller models for L, and H, first an l-model, with L falling slowly, and an h-model, with H falling rapidly. Now, he is asked to consider how H and L would fall together. This demands an integration of h-model and l-model. When he tries to perform the integration he discovers that an integrated l–h-model, which would be at the same time model for P, is impossible. Instead of manipulating formulae (the way a logician would do) Simplicio manipulates more concrete representations enabling him to use his everyday skills and will certainly reach with ease the required conclusion (the interim conclusion that it is incoherent to maintain that speed is proportional to weight). The purely propositional way might have been, on the contrary, closed to Simplicio.

Similarly with Einstein's famous moving train TE. We are invited to imagine an embankment and a moving train passing. We consider two distant points at the train, A and B and the midpoint M between them; at the given moment two lighting strokes strike at the two points.

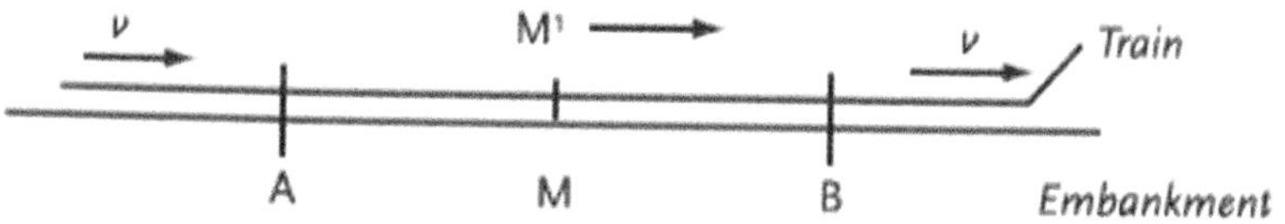

Here is Einstein's description:

> When we say that the lightning strokes A and B are simultaneous with respect to the embankment, we mean: the rays of light emitted at the places A and B, where the lightning occurs, meet each other at the mid-point M of the length A → B of the embankment. But the events A and B also correspond to positions A and B on the train. Let M′ be the mid-point of the distance A → B on the travelling train. Just when the flashes of lightning occur, this point M′ naturally coincides with the point M, but it moves towards the right in the diagram with the velocity v of the train. If an observer sitting in the position M′ in the train did not possess this velocity, then he would remain permanently at M, and the light rays emitted by the flashes of lightning A and B would reach him simultaneously, i.e. they would meet just where he is situated. Now in reality (considered with reference to the railway embankment)

> he is hastening towards the beam of light coming from B, whilst he is riding on ahead of the beam of light coming from A. Hence the observer will see the beam of light emitted from B earlier than he will see that emitted from A. Observers who take the railway train as their reference-body must therefore come to the conclusion that the lightning flash B took place earlier than the lightning flash A. We thus arrive at the important result: Events which are simultaneous with reference to the embankment are not simultaneous with respect to the train, and vice versa (relativity of simultaneity) (Einstein 2002:28).

Now imagine the train in the scenario, I am told. Well, I am supposed to start from the viewpoint of the outside observer. I notice that the light has been emitted simultaneously from the left and from the right lamp, and I know that it will reach the traveler at the same moment. But then, I have to redo the imagining from the perspective from inside the train. Imagine that the two emissions happen at the same time *t* as in the outside frame and that they are simultaneous; now follow the light traveling to the center, where the traveler is situated. How swiftly will they travel from within the train?

And here the inference starts, starting from the knowledge that speed relates time and distance. The train is moving, say, from left to right, so the signal that travels in the same direction will travel by the normal speed of light, *c*, plus the speed of the train; the resulting speed will be bigger than *c*. All this is a normal inference. Then comes the additional empirical information, from Maxwell's theory: there is no speed in the physical world larger than *c*. So, something has to be revised. The rest in pure inference, a contra-intuitive one for a beginner.

Now, my proposal is that the first part of story is essentially dependent on modelling in imagination. The crucial task is changing the viewpoint, from the outside to the inside one. And this is done *in relation to the imagined space*, focused on the singular items, like the train, and the light sources; no general inference is doing the job.

Now we can address the phenomena which pose a challenge for the non-Platonist. Against the claim that armchair physics is"not a case of seeing old empirical data in new ways", the obvious reply is that some TEs do involve seeing old data in the appropriate way, and some involve combining old data in new way. If no one had looked at the given set of data before, or in no one has had any idea of how to "see" them in any systematic way, then our armchair innovator has just seen them in the right way, without seeing them in "new" way, because there was no old way of seeing. Much more important is the possibility of producing new data by manipulating the old ones. So, the manipulability of elements in the model and the mobilizing force of particular instances, more friendly to our ordinary reasoning process than the abstract general arguments, should account for the heuristic value of TEs. The theory of mental models makes correct prediction that a visualizable TE will have more heuristic value that a corresponding propositional alternative, and explains in principle why.

It is plausible that manipulating quasi-spatial models can in heuristic value come very close to actually seeing the movement of pieces manipulated. The manipulation is quasi-experiential in that it yields non-inferential beliefs akin to perceptual ones. This might explain the felt difference between a TE and a piece of regimented propositional reasoning. (I leave the issue of justificatory value for the next section of the paper).

We have already mentioned the ease and quickness of arriving at the conclusion of TE. Remember the formulation offered by J. R. Brown where he stresses that "the jump from data to theory is relatively small" (Brown 1986:4) but can lead very far; however, this does not mean that the theory suggested by successful TEs is accepted without testing—quite the contrary is the case. In fact, In many TEs there are usually two crucial steps, that cry for an explanation, the particular and the generalizing one. At the first stage, a particularized challenge to conceive of a state of affairs involving some property P is raised (a musket ball and a cannon ball being tied together, or a ship the planks of which are successively being removed). Then the person goes through the TE implementing a P-involving particular scenario. The stage ends with intuition-judgment concerning P in the particularized setting. This gives us the first jump: from experiment to particularized judgment. At the second stage there is passage to generalization As we mentioned, in the Aristotelian tradition, it is sometimes called "intuitive induction". We shall address both steps, without always saying it explicitly.

The contrast between the importance of the jump(s) and the ease and quickness of it is of crucial importance. One important challenge it presents concerns the implausibility of the assumption that the thought experimenter uses inductive reasoning. If she must identify and weight alternatives before reaching her conclusion, as canons of inductive procedure demand of her, she will never make his step so gracefully.

A plausible line of explanation deploys the notion of default hierarchies. E.g. a cognitive system can learn e.g. that (usually) when Jane comes to the party, it is fun and thereby store the information about the existence of disposition D (Jane coming to the party, The party being fun) Then, the system might learn that when Jane and her boyfriend come to the party it is not fun, acquiring thereby D (Jane and her boyfriend coming to the party, The party not being fun.). The first disposition is the default one, whereas the second is activated only when the system comes to consider the possibility that Jane's boyfriend is coming to the party. Of course, reasoning with conditionals involves many more complications beyond the two mentioned, so that a complete account of hypothetical reasoning should take all of these into account; however our interest is much more limited. The essential naturalistic point is that the transitions sketched can be easily mechanically implemented in links and networks, and that cognitive scientists suggests that this is actually how our memory is structured. Different assumptions about regularities will result in different models of possible courses of events.

We can now introduce the proposal about quickness and ease of the jump in TE. Start with ordinary reasoning. When you hear someone saying "Jane is coming to the party, so the party…" you probably automatically expect him to end saying "…will be fun". You do not expect Jane to come drugged, or sick, or in a fit of rage, and so on.

The explanation for the speed and ease with which you hit on the "will be fun" alternative is that your knowledge is stratified. The proposition.

When Jane comes to the party, it is fun is your default alternative. Other propositions, like the one mentioning her boyfriend, are possible defeaters arranged in a hierarchy. The finished default hierarchy presumably embodies the results of a long learning process. Here is then the main proposal.

When a reasoner builds her model, she can use such stratified knowledge in order to reach swift and effortless cognitive decisions. The answer lies in the fact that reasoning in mental models mobilizes the resources of highly organized background knowledge—chunked, Consider the two examples, Einstein's and Galileo's. We all know that light normally travels straightly and not in a zig-zag fashion; this is the default assumption, built into our mental model. Tens of such default assumption determine what it will look like. The action is not about them, but about the simultaneity, and there we reason quickly, with the rest being in background.

Similarly, you all know that color does not affect speed. How do you know it? Your knowledge is organized (and stored) in a default way and it might be present implicitly in various scenarios concerning situations of falling. It is difficult to handle outside a concrete context (try to think of all the things that you in fact know to be irrelevant for the speed of a falling body). If one had to write down a list, and to explain why the particular characteristics are on the list, it would be a never ending job. Return briefly to Galileo's experiment, the second and final part—the jump from irrelevance of weight to the positive definite conclusion that all bodies fall with roughly the same speed. How does one exclude factors other than weight? We just noted how it might happen. The speed of free falling object is independent of its color and of its taste. It is independent of the look of the object.

These truths are built into our background knowledge presumably as default assumptions. If we had to make all of them explicit in order to arrive at the conclusion, we would never arrive at it, certainly not within a few seconds. But the simplest (and therefore proper) explanation of our speed and efficiency is not that we see the right explanation written down in the platonic realm, but that truths about what is irrelevant are given as implicit constraints on our model building. They determine the architecture of the model, while remaining rather inconspicuous in the background.

All the irrelevant possibilities (irrelevant from the viewpoint of the inquirer) are excluded thanks to the filtering power of default hierarchies. (There is of course no special reason to think that the background knowledge itself is a priori. The default hierarchies are presumably built up in some kind of inductive procedure, embodying lifelong experience and learning from others.) This way of excluding possibilities offers an interesting answer to the argument-theoretician as well. It is clear that a full argument-version of a typical thought experiment would have to include a lot of premises representing background assumptions, indeed nested premises in which a given more general premise is qualified by those stating possible exceptions of which the thought-experimenter is presumably aware. (Take the amount of background knowledge about ships presupposed in the Ship of Theseus TE: changing planks does not destroy the ship, it can change the color of the ship but should not dramatically change its weight, etc.). Moreover, there is the issue of the "kinematics" of TEs, in which some premises get updated as the reasoning proceeds. It is incredible that even a very clever thought-experimenter goes through all this material in performing her TE. The nice, streamlined arguments from the book just don't correspond to the actual practice.

The proposed line might go a long way towards explaining why the intuitional induction is justified. It is such, because the irrelevant (or seemingly irrelevant) factors have been neutralized and taken care of by our "background knowledge" (taking "knowledge" in the non-philosophical sense of firmly held belief), and its default hierarchies. The significant amount of such background knowledge is being brought to bear on the issue, and this is done imperceptibly, with no manifest effort. So even the name of "intuitional induction" thus seems vindicated: the generalizing jump itself is inductive in an indirect way.

Let us summarize. The assumption that performing a TE consists in constructing a mental model and reasoning in it, explains the nature of the data, points to the skills employed in manipulating the items in the model, and accounts for heuristic fruitfulness and relative ease of the work in the laboratory of the mind. It denies that one has to interact with possibilities in order to get knowledge of them; instead it proposes that models of possibilities are constructed in a combinatorial manner from components, representation of actual items. It postulates no mysterious capacities, unlike the platonic account, and is therefore preferable to it. Interestingly, it is Kant whose account of "construction in intuition" comes closest to the mental model view. Without the theory of reasoning in mental model, the empiricists did not have the necessary instruments at their disposal—non-apriorists like Kuhn talk vaguely or metaphorically about seeing old data in new light, and even propose the extremely implausible view that the thought experimenter must already have had the experience described in the experiment, all this because of the lack of adequate psychological theory. A critic sympathetic to classical empiricism has once remarked against these claims that "this "theory" (that we arrive at possibilities by rearranging items gleaned from experience) is at least as old as Locke" so the empiricist did have the requisite instruments at hand. Well, it depends on how you read them. Locke and Berkeley apparently both took "ideas" to be picture-like, so they had the wherewithal for an account of the imaginative rearranging. But they lacked a clear conception of the rest: how a particular rearranging interacts with more general inductive information, what the format is in which such an information is systematized, and the like. This might explain why it took several centuries before the proposal was even gleaned by authors like Mach, let alone clearly formulated.

We may conclude that a large part of physical TEs can be explained by non-apriorist means. There are, however, some important debts we have not yet paid. First, we should remember that a part of the background knowledge mobilized in model construction and in imaginative 'experimenting' is properly physical, and concerns forces and movements, whereas is properly spatial. Where does this knowledge come from?

The physical part is relatively easy, but not completely trivial. It comes from people's ordinary proto-theories about physical interaction. The cognitive research has uncovered a certain systematicity in such theories, to the extent of justifying the label of 'naive physics for them. Now these theories themselves must derive from some source or other. There is an intense research going on about the origin of naive physics, and it seems that a part of it is innate. The issue is philosophically rather simple, compared to the usual problems with intuitions, although intriguing for cognitive scientist. We shall leave it aside. To conclude with the science TEs, the model account has an additional explanatory advantage over the platonic one—it holds promise of explaining not only the virtues but also the limitations of TEs in a systematic fashion. Brown once raised the following question: why don't we normally have very reliable intuitions about chemistry? A natural answer is that chemical knowledge is not part of our folk theories, and that chemical reactions are not accessible to us to the degree physical reactions are. Therefore, there are no relevant assumptions that a thought experimenter might use. The model view offers a direction for explaining the phenomenon; I wonder whether the Platonist has anything comparable.

Until now we have been considering one kind of mental modelling, namely model building. We now quickly pass to the other kind, mental simulation. Indeed, two interesting sources make their appearance in the study of mental modelling. The first is what we may describe as the 3rd person construction: the constructor uses her resources from memory and her competences to put together in imagination the elements that will result in a "picture" of the scenario. This was out topic in the preceding section. The second, geared to 1st personal (an interpersonal) understanding is mental simulation. If one is asked in the variant of the Trolley problem, whether one would push the Fat man in order to save several potential victims, one probably puts oneself in the shoes of the bystander next to the Fat man, and tries to simulate the feelings and intentions he would have. The two techniques might yield contrasting results: the standard construction process is way more impersonal than the stimulation, which might engage subject's emotion to a much higher degree (see my reply to David for more examples). So, here is the contrast: model building versus modelling through simulation:

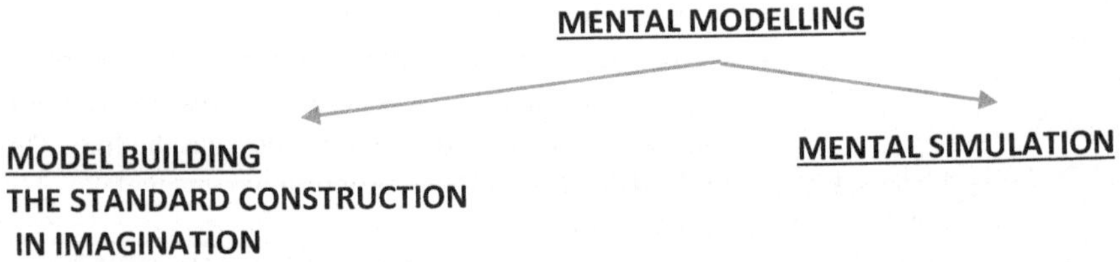

Modelling through simulation, from the 1st person standpoint is geared to human and social matters, but can go wider, as we shall see in the moment, using a famous TE due to Einstein. But first, human matters. In the black-and-white Mary scenario we are invited to imagine what is to be like to know all neurological facts, and then to see, for the first time in one's life, red, green, blue and yellow objects in the outside world. In moral and political TEs one will be modelling human-social relations. In ethics it will be the Ring of Gyges as well as the Trolley and Violinist TEs. In political TEs the subject has to produce a mental model of ideal Platonic state, State of nature and of the Original position and the Veil-of-ignorance.

So, what do subjects model in simulation TEs? In philosophy, for instance, epistemic-cognitive states: in particular ways of forming beliefs.

Note that we are primarily considering the early stages of a TE, from two to five. Take Gettier cases, in particular, the getting the job scenario. Assume that the subject has understood the question, whether Smith knows. In my view, the stage two of tentative conscious production consists of constructing consciously a model of what is happening. The subject imagines, more or less vividly, Smith counting money in Jones's pocket, getting (mis-)information from the boss that Jones will get the job. Most often, the subject also needs help of relevant sources at unconscious level. I have described this as stage three, of unconscious additional production. In particular, she needs the relatively specialized background of understanding of what knowing consists in, and how it is related to other epistemic states. Hopefully, the model building and early search for the answer culminates in the immediate, spontaneous verdict, often non-conscious, our stage four which is the core of the intuition in which the early production culminates. (I leave open for the moment the issue of possible wider unconscious empirical theorizing, our stage four a; I shall come back to it in a moment. For the moment, note that it could encompass pragmatic elements: what does the teacher want from me, should I be nice to Smith or not, how do I comply with what I take to be the teacher's expectations?) The subject finally reaches immediate, conscious spontaneous answer, conscious intuition, say that Smith knows; this was our stage five.

Consider now a particular kind of knowledge, namely the one that consists in understanding[1] Does mental modelling enable understanding and how?[2] Back to Einstein. Suppose I am confronted with his story of the elevator inviting me to do the following. Imagine a person in the elevator, possibly yourself, and imagine that he (as Einstein puts it) has dropped a handkerchief and a watch. No problem, I can simulate the event, put myself into the elevator in my imagination, and follow the handkerchief and watch as they are falling.

Now, imagine something more exotic: "an elevator being uniformly accelerated by a "being" in a gravitational-free space." Einstein and Infeld 1938, p. 231). And imagine a discussion between the "inside observer" and the "outside observer":

The outside observer notes that the lift moves upwards with constant acceleration, because a constant force is acting. This I can model in my mind: just putting together a picture of the elevator being constantly accelerated. In the model the following happens: "If a body is left free, it soon collides with the floor of the lift, since the floor moves upward toward the body [...]" Ibid. Call it the model One.

The inside observer has a different impression: "My watch, my handkerchief, and all bodies are falling because the whole lift is in a gravitational field." (Einstein and

---

[1] leaving aside the thorny issue if there are cases of understanding that are not cases of knowledge (see papers in Grimm et al.; for general information a good introduction is Chap. 1. of the book, entitled "What is understanding?" by Christoph Baumberger, Claus Beisbart, and Georg Brun).

[2] I was inspired to stress understanding in TEs through reading of M. Stuarts excellent "How thought experiments increase understanding?" (Stuart 2018). See also his papers on imagination (Stuart 2019) and (Stuart 2020).

Infeld 1938, p. 231) My model contains a person, standing in the elevator: his legs are under the pressure of the elevator floor, just as they would be if the elevator were in the free fall. He observes his watch coming close to the floor and interprets this as watch falling. Call it the model Two.

And suddenly, I realize the equivalence: "One cannot differentiate, by observation, the fall of bodies in a uniformly accelerated referential frame away from a gravitational field, from the fall of these same bodies in a frame at rest in a gravitational field." (Ibid.). I *understand* that the two are indistinguishable. Model One and model Two depict the same situation. And this understanding, derived from *equivalence of my two distinct models*, can take me very far. It suggests that the two frames are equivalent: uniform acceleration away from a gravitational field is equivalent to the situation of the frame at rest, but in a gravitational field. Thus, Einstein postulates the equivalence principle between a uniformly accelerating frame with inertial field and the corresponding un-accelerated frame with a homogeneous gravitational field.

The link between modelling and understanding should be developed further and it is an important task for the future research on TEs.

4.2 Intuitions: Moderate Voice of Competence (MoVoC) Theory

What abilities underlie ordinary model-building, in particular in philosophical and scientific model-building? Probably, some general abilities, like intelligence and memory. Next, we need ability to handle counterfactual situations, which is specialized in its function, but general in regard to specific domains. Finally, we need specialized ones, probably quite a bunch of them.

For physical TEs one needs spatial competences, and ability to manipulate naïve physics (take Galileo's Free fall TE as a central example). Now, for the Ship of Theseus and Brain switching, one needs general understanding of identity in time, change and persistence, and this might be a part of our most general understanding of the world, of ourselves etc., the "naïve metaphysics" of ordinary material object, persons and the rest. My own favorite line on this set of competences is summarized in the following paraphrase of J. L. Austin (where is term "words" is replaced with "concepts"): the stock of our deepest concepts embodies all the distinctions men have found worth drawing, and the connections they have found worth making, in the lifetimes of many generations: these surely are likely to be more numerous, more sound, since they have stood up to the long test of the survival of the fittest, and more subtle, at least in all ordinary and reasonably practical matters, than any that you or I are likely to think up on the spot (1970/1962, p. 182). This accumulated wisdom then allows the philosopher to anticipate the experience from the armchair. At the same time, the double fallibility of these intuitions accounts for the limits of philosophical autonomy: armchair research should be open to corrections from empirical science.

Also, our innate endowment might explain at least the very origin of the intuition-capacity and the initial stages of the formation of our intuition-states with their contents. For instance, it might consist of innate structures, some corresponding to concepts and some to inner, spatio-temporal "frames", responsible for an innate spatial-geometrical know-how to which Devitt is alluding in his considerations from analogy. It explains some of the objective validity of our intuitions. But nativism should be restricted to the origin of the system and to the relatively initial stages of processing. An intelligent nativist-adaptationist should allow for a wide margin of influence from individual empirical learning, which may overthrow even some deeply ingrained pre-conceptions to the contrary. And most importantly, intuition is doubly fallible. It can misrepresent the contents of our cognitive apparatus, and is thus internally fallible. But, the contents themselves–including their core, the innate assumptions–are also fallible, yielding the external fallibility of intuitions. Our innate geometry might be false, out possibly innate folk-physics certainly is.

Consider now norm-and-value related abilities. With epistemic-cognitive states we need a normative background of evaluating ways of forming beliefs, competence in recognizing epistemic statuses, especially knowledge. For modelling human-social relations we need specialized competences for understanding, model-building and simulation. More importantly, for moral judgments, we need our moral competence, whatever it might be, moral grammar, or moral sentiment, or some third ability. Whatever it is, it is highly specialized.

Where do these competences, crucial for intuitions-judgments in TEs, come from? It's time for some polemics. Consider intuitions about one's mother tongue. Remember the sentence:

(1) John seems to Bill to want to help himself

The naïve speaker of English immediately recognizes that it is John who appears to want to help himself. How does she manage to do it? On the standard Chomskyan account her competence, embodied in the parser in her head produces a structural description, a tree where "himself" is tied to John; this information enables the hearer to understand who is the "himself", the object of the alleged intended helping. The hearer also understand that it is not Bill who wants to do it, rather that Bill is confronted with this seeming about John, and what John wants. And the like for tens of thousands sentences. The competence is partly innate, and only this fact can explain that naïve speakers can produce competent judgments about syntactically complex sentences like the above one.

Now, the idea of such special competence, not totally dependent on empirical learning has criticized by some authors. Take as example linguistic intuitions, where the competence view has been criticized by Michael Devitt (2006); we shall assume that similar arguments for and against competentialism can be made about other competences, for example spatial-geometrical and numerical ones. Take the example of a native speaker confronted with a sentence like *'Tom thinks Dick loves himself'*. Whom does Tom think Dick loves, Tom or Dick, she is asked, and she replies "Dick". Correct grammatical intuitions. Where do they come from?

Our intuitions and abilities to generate them are not special, the critic says, they are just result of long empirical learning, the critic claims. They are analogous to impressive capacities that specialist have on the basis of long training and activity. Devitt offers the example of a paleontologist responding to a bit of white stone sticking through grey rock with "a pig's jawbone"; next he borrows from Gladwell (2005) other nice examples: of art experts correctly judging an allegedly sixth-century Greek marble statue to be a fake; of the tennis coach, Vic Braden, correctly judging a serve to be a fault before the ball hits the ground. A person's "linguistic intuitions are *perceptual* judgments", he claims (2014:14).

What can the competentialist reply? She can say, first, that they feel like perceptions. To return to Tom and Dick example, the competent hearer normally 'hears" the sentence as claiming that what seems to Tom is that Dick loves himself, Dick. Immediate understanding of sentences in one's mother tongue is very close to perception; it follows immediately after "hearing" in the very narrow sense of the term, call it hearing-n, and thus as completing the "full hearing", call it hearing-f. The question now concerns the role of competence in such hearing-f. Devitt just appeals to "a speaker's years of experience and education" that allegedly prepare her for such hearing. But can mere experience with language teach one the complex syntactic structure of ordinary sentences like the Tom and Dick one? Can it provide the representation that has this relevant feature (himself = Dick). The recognition must depend on a relatively complex tree-like structure somewhere in the mind; it is probably not present in the central processor (too specialized for it); so where else can it be but in the competence-parser?

Second, Devitt admits that general intelligence (the "central processor" does not have "the informational basis for the intuitive judgment that 'himself' co-refers with 'Dick'; rather it is the experience that supplies the intelligence here. So, what does have it? The speaker comes up with the intuition and the intuition is correct. Where does it come from, if the general intelligence does not have the informational basis? Where from, if not from competence-parser?

Finally, let us return to the examples of experts. I agree that there is an analogy: The three heroes, paleontologist, art expert, and Vic Braden are experts in their respective areas. I, a naïve Croatian speaker, might be as good in recognizing and understanding the sentences of my language as an art expert is good at recognizing fakes. But there is difference in detail. How do you know this piece of art is a fake? Well, here are my reasons, the expert hopefully answers, and lists some sophisticated indications for this judgment. His judgment is based on expertize, acquired by academic study, with a lot of explicit material, all worked out by his CP, stored in his memory and accessible to his consciousness. My naïve linguistic judgment is very different: I cannot draw for you the parsing tree for our example (nor for its Croatian equivalent). I have no memories of having learned the parsing trees for relevant language(s). The informational basis for my judgment is not there, in the field of my general intelligence. It has to be somewhere else. Why not in the parser? This is what MoVoC suggests.

It is not difficult to widen this kind of reasoning and apply it to competencies dealing with space or number; for specific deeper problems with mathematical domain see the Appendix. Let me add that there are specialized failures in the spatial perception: experts talk about disorders of visual space perception associated with lesions of the right cerebral hemisphere and the like. This adds to the assumption that some basic features of the competence are innate, and that the ability itself is not just the result of empirical learning. Of course, much is learned, and competences are honed in practice.

Now that we have a picture of how deep the competence basis goes, we can understand the need to distinguish the competence-driven model-building from possible interventions of general interests and capacities. Motivation, general intelligence and knowledge do interfere, helping or spoiling the results yielded by specialized competence(s). As far as my moral capacities go, I might be very much in favor of throwing the Fat man down from the bridge, but I might feel bad about telling this to my class; I deceive myself, and on the conscious level I produce the (somewhat hypocritical intuition) that I would not do it, and that it is deeply wrong to do it. The contrast is very much needed for issues in X-Phi, see the next Chapter, where an explanation of intuitions is the most important step in selecting the valid ones.

We have thus got several important topics, concerning imaginary enactment in general, model building, thought-experimenting and intuition-producing fitting together. So, we now have the minimal necessary elements to formulate a proposal concerning the nature of intuitions and TEs producing them. I have called it Moderate voice of competence view ("MoVoC" for short). It starts from the admission that there are intuitions-dispositions and judgments, which form a distinct group of phenomena, and there is the intuition-capacity, the capacity to use our imaginative and judgmental competencies in an off-line fashion. It is the voice of competence, most often discreet. Intuitional data are thus the minimal "products" of tentative production –linguistic, philosophical, moral or mathematical– by naïve thinker (or speaker-listener) and not their opinions about the data. The data involve no theory and very little proto-theory. Although there might be admixtures of guesswork in the conscious production of data, these are routinely weaned out by linguists. As against predominantly conceptualist understanding of TEs and intuitions (Peacocke, Boghossian) it claims that intuitions are concerned with their external objects, the domain of items and facts, rather than with concepts. Concepts often play a role in the process, but they are not the object of intuitions, and their role is subordinate to the role played by the external referential domain.

Finally, concerning TEs: they and the intuitions they produce are rather scenario-based than inference-based. Imagining or simulating scenarios, typically particularized ones play the main cognitive and justificatory role, whereas inference typically plays a subordinate role. Fourth, explanationism: intuitions require an explanation of having and reliability, if possible a causal one.

It is probable that intuition is not unitary and that the relevant capacities are very diversified, used in a wide range of non-intuitional tasks. What is common to all intuitions is more their "manner of presentation" than the underlying structures. Primitive intuition is quite fallible, and a lot of ulterior work is needed to filter out the valid intuitions from the non-valid ones. The sophisticated intuitions that characterize the work of mature scientists involve a lot of learned skills and information. The evolutionary, adaptationist hypothesis offers a hope that a part of our primitive intuitional knowledge does reflect the deep make up of our environment, and thus, in spite of its fallibility, carry information about real and philosophically important properties of some states of affairs in the world–say the mathematical ones–prominently about their necessity and universality (see the Appendix for more).

This is then a short summary of the mental modelling proposal and the moderate Voice-of-Competence view that goes with it. They are naturalistic, and turned to cognitive-scientific explanation of thought-experimenting, and they put in the center the elements of imagination-based accounts, but they leave some space to inference and computation.

4.3 Appendix: The Puzzles of Geometrical and Numerical Intuitions

Number and geometrical properties pose deep epistemological and metaphysical problems, which we can only very briefly sketch here (I develop my own view in Miscevic.

Let us start geometrical examples, and to the target claim that by folding a square over one of its diagonals, the opposite sides will come to coincide. The mental model account suggests that here again the cognitive system uses the knowledge about the spatial properties of the perceptual world normally needed in order to process the incoming stimuli. When one imagines a geometrical configuration, one's imagery is being directed by perceptual-imaginative sub-systems, that process the awareness of real shapes in real objects. The serious imagination is controlled by habits and routines of perception. In our example, when one imagines the angle growing bigger, one's internal imaginary movie is being directed by perceptual-imaginative sub-systems, that process the awareness of real angles in real (approximate) triangles. The intuitions of geometrical possibilities derive from constraints imposed by the quasi-perceptual origin of human imagination. If the laws of visual geometry are written down in our brain, people can discover them by careful imagining. (Of course, it is easy to supplement this explanation by assuming that other imagined contents come from experience and combine with the hardwired geometrical organization). Let me then end with reverting to the structure we encountered before, and encounter again here: a possibly hardwired set of representations controlling both behavior and intuition.

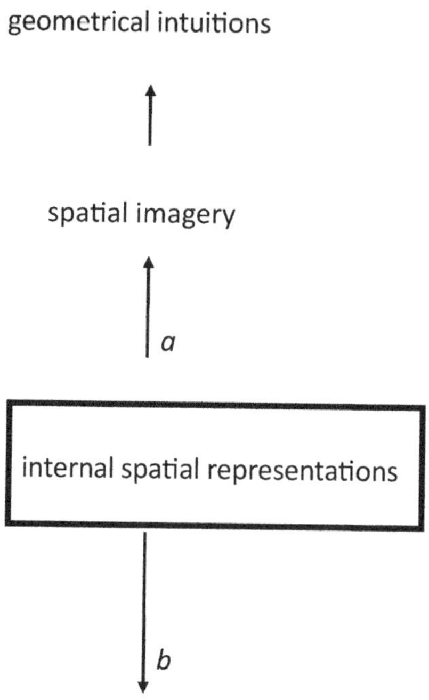

The internal representations are probably partly innate and partly acquired by learning.

Now, as we mentioned before, the visualized sides are not ideal lines, nor are they necessarily represented by imagined lies of absolutely exactly equal length. It is here that the anti-visualists enter their disagreement. Let me quote M. Giaquinto, one of the most sophisticated and cautions contemporary critics of the view that visual presentation can justify beliefs in theorems of geometry: "We have no reason to believe that such features instantiate perfect geometrical forms, no reason to think that squares are represented by perfectly square images" (Giaquinto 1992: 387). (As Mike Martin has put it in a discussion, imagined triangles do not seem to be more precise than the ones drawn on the blackboard, and those are admittedly to imprecise to buttress by themselves a proof). Let me put the objection in a more perspicuous form: The (naive) cognizer rationally thinks that her geometrical beliefs concern perfect geometrical forms. Visualizing experience does not offer perfect geometrical representations. Therefore, visualizing experience cannot be (rational) evidence for the geometrical belief in question. Epistemological Platonists, perhaps from Plato himself all the way to Meinong and J. R. Brown have built upon this objection their view that imagination is just a heuristic aid to a reason-based faculty. Others seek refuge in non-intuitional formal foundations of geometry, forgetting that intuition often serves as a guide to formal approaches–as their initiator David Hilbert himself has been stressing–and has to be accounted for nor matter what ulterior method mathematicians have at their disposal.

To the visualist the consideration of the absolute precision counsels modesty: the evidence from visualizing is certainly only a part of the story, not the whole story. To see why consider first the radical visualist who wants to put precision into the images themselves. There are two possibilities open to him, but none of them sounds very impressive, and none of them is in any way confirmed. Let me expound them using the lines example. The first possibility is to assume that already visualization as well as in perception one often does not see the lines of any definite breadth. The visual "lines" are sufficiently thin to serve as a model for geometric lines. The second possibility is that imagination is more schematic than perception. We imagine geometrical types, so to speak. Interestingly, some cognitive psychologists talk about "schematic images", that seem roughly to yield what is needed, i.e. a kind of image that "abstracts" from the inaccuracy of detail.

Pending future research, the best line is to be modest and propose a non-radical option: the visualizer does not imagine perfect geometrical lines, but only the thick and imperfect quasi-lines, and only at the next stage arrives at geometrical lines by inference, which might include abstraction and idealization. The right description of what happens in our second example is then the following. By "folding" the imagined square the naive cognizer can "see" from visualizing that the two sides c and d will be aligned and will roughly coincide, then mobilize the conceptual information that they are equal, and infer, that, if they coincide, they will coincide exactly. But doesn't her progress reduce the role of "seeing" to a mere reminder, which becomes evidentially superfluous once the conceptual information is mobilized? (Pierre Jacob has voiced the worry in a discussion in Lyon) No, since the concept mobilized does not need to be that of rotation and alignment; the equality is enough. Therefore, the corresponding first step concerning the "folding" can be supplied by the imagination. The concept takes care only of the non-imaginable absolute equality. Some anti-visualists, for instance Giaquinto (1992: 395), admit that our geometrical concepts have to be accompanied by "concept images"–presenting the transformations of geometrical items–in order to support geometrical reasoning; now such images seem to encapsulate precisely the visual aspect that is crucially needed for naive geometrical knowledge. Notice also that in the example, the actual breadth of the sides plays no role whatsoever in establishing the desired conclusion. In this case, then, one can take the traditional empiricist line and claim that the naive cognizer abstracts from the actual breadth of the line-like item, and that she is justified in doing so The abstracting might be made easier by the fact that the line-like item is usually not ascribed any particular breadth by the perception or imagination themselves, not represented as having this or that particular breadth.

This brings us back to Benacerraf's dilemma. Remember that its first horn is realism about numbers. The realism gives you nice, literal truth conditions for sentences like "Two plus three equal five", referring to numbers. But then, the problem of knowledge raises its ugly head: how can we know the truth about such abstract objects? The Second-horn: anti-realism about numbers: numbers are our mental constructions. Then we have a lot of other problems, e.g. that it sound unnatural, that it is mysterious why mathematic is successful in natural science, etc.

Here are the philosophical assumptions I would like to keep. First, that there are real (mathematical) features of the world, assumption shared by most realists, and impressively argued for by structuralists. Second, that elementary mathematics captures some real (mathematical) features of the world, formulated by classics like Aristotle (1988) and nicely accounted for by structuralists. Third, that elementary mathematical intuitions ("Two plus three equal five") are realistically correct. Fourth, that elementary mathematical intuitions are cognitively first in the order of time, individual and historical. (I would also like to argue that they are epistemically basic in a more foundational ways, but I cannot do it here). We do intuit the concrete mathematical items (instantiated patterns, pairs, trios, quartets of things of all sorts), and nevertheless, as Frege would put it seeing what number is "as if through a mist". How is this possible?

The problem is simple but deep. Of course, we do perceive objects and our perception concentrates upon similarities in color, overall shape and Gestalt-organization, and, when faced with a choice, disregards more abstract similarities like the numerical one. Therefore, as against the structuralist mainstream, Parsons and Resnik in particular, *no purely perceptual device would do for recognizing instantiated (structural) mathematical properties.* Typically perceptual properties like shape, color, gestalt-features in general are manifestly unsuitable. The conceptualist critics then often suggest that the number concept is the only one which would. Intuition would then be demoted from its role of providing the first contact with numerical properties. We hope to have found a middle road, with the hint from the cognitive research: *simple, pre-conceptual, or, at most proto-conceptual operations are sufficient to ensure the right initial classification (typing) of particular items into numerically relevant types.* Further, the operations might at the outset be performed by purely unconscious, mechanical, image-based or proto-computational routines. Their initial result might be that the beginner (a child) treats sequences as if they were classified by their proto-numerical properties. Later on, part of the procedure–or even the procedure as a whole might become conscious, r even an object of reflective attention and deliberation, but retaining the same abstract structure, the very scaffolding which filters out the right properties. we reason to the non-instantiated ones. We thus have indirect contact with "original" mathematical universals, and a naturalistically explainable competence.

The target of elementary mathematical activity are objects and object-properties, things in the outside world, their collections and the numerical properties of both. This suggests the explanation of the utility of having counting-and-calculation routines available. Thus we need not limit ourselves to present interactions: computational abilities have evolved through evolutionary history and the humans have inherited the fruits of evolution. The intermediate target might be "interior" objects, mental marks, mental images or other representations. Mental operations—prominently including visualizing, imaginative ones–performed upon appropriate items are thus the instrument, but not the proper object of mathematical cognition, as Kitcher, for instance, would have it.

Remember that the definition of equinumerosity in terms of matching (1–1 correlation) suggests that one can establish it by imaginatively matching the elements of two or more collections. Now if basic imaginative intuitions are really fundamental for acquiring numerical knowledge, and if that knowledge does concern abstract properties, it follows that imagination can reveal abstract properties. If this holds, an account based on such assumption might (and I hope would) explain in a non-question begging way how the abstractness of the target properties is compatible with their quasi-perceptual detection. But how can abstract numerical properties be thus detected?

Structuralism (in versions offered by Parsons, Resnik and Shapiro, see References) offers us a plausible metaphysics to be combined with the epistemic message from cognitive science.[3] In a nutshell, the idea is that *the initial degree of abstractness is achieved through the choice of the right operation: being insensitive to many concrete properties–color, size, and the like–the matching detects the abstract structural property of (equi-)numerosity.* This goes a long way towards solving the problem of 'perceiving' concrete objects as instances of abstract number-structure. Of course, the proposal is also compatible with less Platonic views than the ones of mainstream structuralists just mentioned; for instance, it is compatible with Neo-aristotelian views, and with even more modest views of mathematical objects; however its strength lies in its compatibility with ambitious structuralist Platonist proposals.

Of course, there is a long road from empirical items to the postulated platonic objects. The initial stages concer more concrete patterns, where the naive cognizers intuit the concrete correlation and "see" that the two pluralities have something in common. What then follows is proto-theorizing and mature theorizing about the underlying abstract structure. One might complain that simple intuitions concerning number and primitive arithmetical operations, our baby-arithmetic, fall outside mathematics and are irrelevant for mathematicians trying to understand their discipline. If by "mathematics" one means professional mathematics. I can do little here to alleviate the worry, since the discussion of relation between visual arithmetic and the 'serious' math would just take us too far afield; I will mention one optimistic possibility nevertheless. Perhaps primitive numerical intuitions and the professional advanced mathematics are just two separate stages of a single cognitive enterprise. The future mathematician starts, as a child with the former, and progresses towards the later. The history of mathematics has taken the same course since its Babylonian and Greek origins.[4]

The point suggests a division of mathematical intuition-states (and -contents) that looks anodyne, but, if valid has far-reaching consequences. One should account first for simple intuitions, concerning finite, surveyable collections and second, for advanced views on sets, infinity etc. The elementary intuitions–those at the first

---

[3] Stewart Shapiro has put the point nicely: "My account of the relationship between mathematics and science begins with the suggestion that the contents of the non-mathematical universe exhibit underlying mathematical structure in their interrelations and interactions. For example, it might be claimed that a mathematical structure similar to the inverse-square variation of real numbers is exemplified in the mutual attraction of physical objects. In general, physical laws expressed in mathematical terms can be construed as proposals that a certain mathematically defined structure is exemplified in a particular area of physical reality" (1983:538).

[4] See, for instance the work of Netz (1999)

stage–provide contact with concrete items and their collections–real or imagined. Being linked to such items they are also testable on them. The simple contents they feature–two items plus three items equal four item, and the like–are subsequently used and radically reinterpreted in the second stage when the cognizer builds theories, elementary ones (proto-theories) at first, then increasingly sophisticated ones, which systematize the data from the first stage. Some obvious-looking theorems are discarded if they collide with most other obvious looking-ones. The result is a coherent whole, a semi-organized theory. Very often the theorems of such a theory are intuitively obvious, and the axioms are accepted because such theorems can be derived from them. Finally, the mathematician axiomatizes the domain, and the semi-organized theory turns into a fully organized, axiomatic one.

We can now return to the easier part, shapes and spatial skills. Whereas Brown thinks that intuition capacity is a basic capacity I prefer to think of it is a derived capacity that employs various basic capacities, prominently reasoning and quasi-perceptual imagination in the off-line fashion. From the point of view of researcher into cognitive science and descriptive epistemology, it offers a "second window" to those basic capacities. We can now generalize our diagram concerning spatial skills. A lot of our knowledge–both knowledge of propositions and knowledge-how involved in procedures and operations–is structured with the help of deep cognitive structures, some of which are innate and only partially sensitive to individual learning. One can read off the working of such structures indirectly, by analyzing the edifice of human knowledge, using it a an indirect window upon the human mind. Primitive intuitions offer another, more direct window upon the same structures. The structures themselves concern various domains–space, time, physical interaction, number, most general properties of things, logical operations–and have evolved in interaction with environment. They have been studied by cognitive science, and some initial results have been obtained. Here is the generalized diagram:

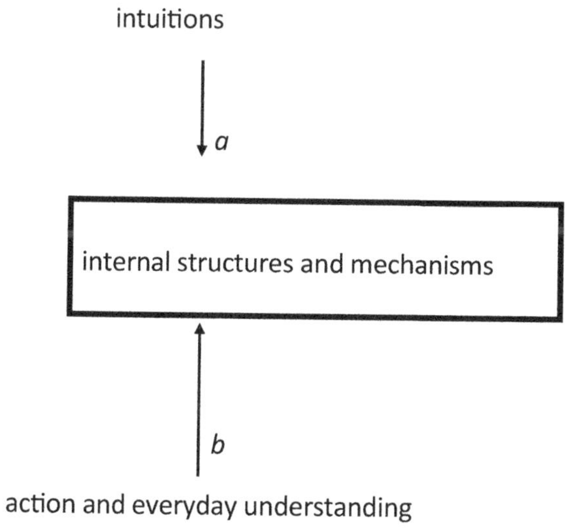

Intuitions should be studied as any other sources of cognition; one should search for already known capacities and try to account for intuitions starting from them, instead of ad hoc postulating new capacities.

References

Aristotle M, Books MN (1988) Translated and commented by Annas. J, Oxford: Clarendon Press

Brown JR (1986) Thought experiments since the scientific revolution. Int Stud Philos Sci 1(1):119–128

Einstein A, Infeld L (1938) Evolution of physics. Cambridge University Press, Cambridge

Einstein A (2002) Relativity the special and the general theory. (originally 1920, London: Methuen). Routledge, London

Giaquinto M (1992) Visualizing as a means of geometrical discovery. Mind Lang 7(4):383–401

Gladwell M (2005) Blink: the power of thinking without thinking. Little, Brown, New York

Netz R (1999) The shaping of deduction in Greek mathematics a study in cognitive history. Cambridge University Press, Cambridge

Shapiro S (1983) Mathematics and reality. Phil Sci 50(4):523–548

Stuart MT (2018) How thought experiments increase understanding. In: Stuart M, Fehige Y, Brown JR (eds) Routledge companion to thought experiments. Routledge, London, pp 526–545

Stuart MT (2020a) The material theory of induction and the epistemology of thought experiments. Stud History Philos Sci Part A

Stuart MT (2020b) The productive anarchy of scientific imagination. Philos Sci 87:968–978

Chapter 5
Do TEs Have a Life of Their Own?

(Understanding the History of TEs)

5.1 Introduction

Do thought experiments have a life of their own? Ian Hacking raised the question almost three ago, and answered it in the negative. A TEs is like a script repeated by various actors, but not further developed (1992:307). They are static. They become fixed, he claims, more like jokes or optical illusions (Ibid.). Unlike real experiments, they have no life of their own. Others disagree. The recent Routledge Companion to TEs (Stuart et al. 2018) notes in the Introduction that "the history of thought experiments shows that many have acquired lives of their own, despite Ian Hacking's (1992) claim to the contrary. This is an important datum for developing ontologies and taxonomies of thought experiment" (2018:3) and they talk about "the interdependence of the history and philosophy of thought experiments (Ibid.).

In this chapter I want to suggest that the attention to the internal structure and stages of a TE discussed through time might help understand its life and systematize its history. Recently the publication of decades long discussions of TE is becoming a routine. Andrew Pessin and Sanford Goldberg have collected important papers on Putnam's Twin Earth TE in their *The Twin earth chronicles* from 1996. Two years ago Cambridge University Press has started a series, Classic Philosophical Arguments, pretty much concentrated on TEs, like The Brain in a vat (Goldberg (ed.) 2016), the Prisoner's dilemma and the Original position. Oxford University Press has published a collection on Gettier problem in 2017 (Borges et al. 2017).

The lives of the two longest discussed TEs, Descartes Evil Demon in theoretical philosophy and Social contract in political philosophy have been amply documented by various published volumes; for Descartes, the classic is Hooker's (1978). TEs in science offer a similar picture. Maxwell's Demon has been discussed since its birth, in 1871 to our own days; the most critical contribution is Norton's extremely negative judgment from 2018. The quantum-physics TEs are filling various collections with a spate comments and discussions.

Given the enormous quantity of material, we have to choose in order to make the approach to be proposed here clear and understandable. The choice will be Hillary

N. Miscevic, *Thought Experiments*,
https://doi.org/10.1007/978-3-030-81082-5_5

Putnam. He is one of the most successful thought-experimenters in twentieth-century philosophy; he has produced two great thought experiments, Twin Earth one and the Brain-in-a-vat one, and many related smaller ones.[1] The two great ones have marked analytic philosophy of our time, from philosophy of language through philosophy of mind to philosophy of science and general metaphysics. Here we shall look at his Twin Earth thought experiment (Twin Earth TE for short), anticipated by him in his 1970 paper "Is semantics possible?" and formulated in 1975 in "The Meaning of 'Meaning'", and at a recent criticism of it, due to Peter Unger (2014).

To illustrate the confusion that the richness of discussion, combined with a lack of clear methodology for dealing with it, we shall briefly look at Unger's complaint about endless "spates of papers" generated by TEs in recent analytic philosophy. We shall argue that the prolonged discussions of fundamental TEs. TEs in science and in various (sub-) disciplines of philosophy define big chunks of the history of science and of philosophy respectively (think of three centuries of debates over Social contract TEs and Evil Demon TE). The chunks are normally organized in recognizable trails. The meta-philosophy and epistemology of TEs has to take this phenomenon into account, thus connecting history of philosophy with meta-philosophical theorizing. We shall argue for this suggestion in connection with Twin Earth TE, and the ensuing debate.

Let me reiterate. The meta-philosophy (or epistemology) of TEs has to join on the theoretical side, connecting history of philosophy with meta-philosophical theorizing. This chapter discusses the meta-philosophy and methodology of the history of thought-experiments, i.e. their neglected diachronic aspect, focusing on Putnam's work as the paradigm case, and on the trail(s) that developed out of the Twin Earth thought-experiment. Putnam's experiment is discussed from a perspective that combines meta-philosophy and actual history of analytic philosophy. Peter Unger has branded the whole debate around it as empty and fruitless. A meta-philosophical analysis shows him to be wrong. The experiment as originally proposed already appeals to a broad variation of examples and intuitive induction; the variation continues in other works addressing the issue, and produces a trail, rich with interesting results. The second aspect is the search for reflective equilibrium, lasting till the present day. The internal logic of these processes is discussed, in order to show that the accusation for emptiness turns against Unger himself. In general, debates around thought-experiments, the already famous and also around new ones, make a large part of contemporary analytic philosophy. The way to understand a large part of this, and of debates surrounding it, is to link it to the internal understanding of a typical thought-experiment; stages of a particular experiment get discussed, developed and changed in the history of a particular trail produced by it. This is an important way in which a philosophical tradition is born, and we need to combine synchronic and diachronic approaches in order to understand it.

Here is then the preview. Section 2 focuses on Twin Earth thought experiment, starting with a short reminder, and continuing with a brief proposal of how to

---

[1] Thanks go to Hilary Putnam for his discussion with me on Skype long time ago, to Smiljana Gartner and Majda Trobok.

distinguish stages in thought-experimenting, in particular within the Twin Earth example(s). Section 3 passes to Unger's criticism and then turns to the defense of Twin Earth TE. Section 4 continues the discussion with Unger by turning to the trails-traditions that have sprung out Twin Earth TE, and connecting the trails to various aspects of the experiment itself. Thus the short-term and long-term patterns of Putnam's thought-experimental reasoning and its continuation are brought together. Section 6 attempts to generalize the morals of the defense beyond Twin Earth cases, first mentioning Putnam's other famous TE, the Brain-in-the-Vat, and then moving very briefly to other examples central in the history of philosophy. It then summarizes our answer to one of the central questions of the chapter: why go historical in the meta-philosophy of TEs, rather than stay with the usual style of topical discussion?

5.2 Reminder: The Twin Earth Te and Its Stages

In order to start proposing a sketch for systematizing the histories of various TEs, based on the typical internal structure of each, let me remind you of our general characterization of thought experiments. A TE normally involves.

1. the formulation of experimental design pointing to
   (i) the determination of the goal(s), in particular the thesis/theory to be tested, and
   (ii) the construction of a (typically) counterfactual scenario to be considered
2. the presentation of the scenario thus constructed to the experimental subject (either the author of the scenario herself, or an interlocutor),
3. the (typically imaginative) contemplation of the scenario and some piece of reasoning,
4. the decision ("intuition") concerning the thesis/theory to be tested.

And then the variations and genralizations from the result. 4. the decision ("intuition") concerning the thesis/theory to be tested. Once this result is achieved, it can be, and often is compared with results of other thought experiments in the vicinity.

In his "The Meaning of 'Meaning'" Putnam famously introduced his Twin Earth TE as a "science-fiction example". We quoted it in chapter Four, so let me just briefly remind the reader. One is invited to suppose that somewhere in the galaxy there is a planet very much like Earth, to be called 'Twin Earth" (Putnam 1975:223). However, the liquid known there as "water" and sharing manifest properties with our water, contains XYZ and *not water* (Ibid.) Let me note that I have asked Putnam in conversation whether we can use Kant's idea of "separating" in "experiments with our mind" of element that normally go together to characterize what he is doing in Twin Earth scenario, and he very emphatically agreed.

Next comes the question for the reader: when a Twin Earthling uses the word "water" does he refer to water? So, after the stage of presentation and question raising, we presumably have reader's understanding. The reader is expected to imagine the situation, to build a model, so to speak. The next comes the reader's intuitive answer. The expected answer is negative: the Twin Earthling refers to what is in fact XYZ, and it is not water!

Let me mention some obvious problems with the performance of the TE at these early stages. First, consider the construction of a (typically) counterfactual scenario to be considered, and its presentation at the stage one. The proponent, say Putnam, is testing people's intuitions, and hopefully, the non-specialists will be included, since he wants ordinary intuitions from non-experts (as well as from his colleagues, the experts). So, the story has to be *relatively short and not too technical*; the best would be if it were not technical at all. This is the first sore point: any expert will be able to come with some counterexample to the short story. Note the parallel with "real" experiments: typically a single experiment concerns just one variable, and taken in isolation proves very little. What is being done is rich controlled variation; real laboratory is in this respect very similar to the laboratory of the mind:

> One feature of experiments closely connected to learning is their variability: scientists perform them over and over again with modifications which may be systematic and intentional or intuitive and exploratory. Variation is one of the key factors in the success of experiment. Frequency of variation in Faraday's experiments is apparent from his laboratory notebooks: these records suggest a lot of unrecorded and unpremeditated variation. (Gooding 1989:67)

And here is the simplest illustration, in the passage we just quoted: "I shall suppose that the oceans and lakes and seas of Twin Earth contain XYZ and not water…", Putnam wrote at the very beginning of his presentation of Twin Earth. But does this not assume that XYZ is not water? What if the reader has the opposite feeling: if it looks like water, tastes like water, and so on, than it is water; I don't care for chemistry! Too bad, the TE was supposed to test intuitions, not to suggest them!

The construction and the presentation culminate with a question. We might assume that the main question in the original version of the TE concerns reference: does the word "water" in the Twin Earth language refer to water or not?

Next, the understanding. It is happening within the conversation, or at least the reading. Strong pragmatic considerations might intervene. First, the motivational ones. The recipient might wonder why she is being asked a particular question, what are the expectations, and so on. Next comes the understanding of the content received; no non-technical, ordinary piece of discourse is free of potential ambiguity, and we know these days that ambiguities often do show their teeth.

As noted in chapter Two, I shall assume that the recipient next builds her model of the situation; let it be the stage Three of the initial TE. This stage is not controllable by the proponent any more, and the model-building depends a lot on specific skills, memories and interest of the recipient. But Putnam does suggest the answer, although he should not be doing this. Let me assume that the recipient goes through some unconscious reasoning; we shall not need it here. So much about the problems.

The stage Four is then the rise of intuition concerning the thesis/theory to be tested.: "I seems to me, very obviously, that the Twin Earthling does not refer to water".

Normally, the recipient is expected to generalize: is what is valid for water also valid for gold, silver, and so on? This intuitive induction ends with a general judgment; in this case "what is valid for water should be valid for gold, silver, and so on." I would add some obvious and immediate consequences, for instance: it is the deep structure of the matter that ultimately determines reference for typical mass terms.

Again, this stage, our number Five, is rife with problems. Assume that the intuition is valid for gold; what about "wine" or "brandy"? Is it really the chemical composition that counts, rather than drinkability, taste and the effects on the consumer?

What normally follows is theorizing, the search for reflective equilibrium. Putnam did find one, and a quite radical one, for that matter, namely a restructuring of the whole of semantics. This is the final stage, number sever, the reflective equilibrium itself. We shall return to it later.

So much about the first (version of) Twin Earth TE. We noted the existence of challenging points at each stage. If you are pessimistic, you can call them "sensitive points" of a TE, if you are optimistic, you can describe them as "productive points", since they obviously invite questions and discussion.

Why "productive"? Consider how the experimenting proceeds further, in Putnam's text. After coming to the conclusion, one might ask a slightly different question, introducing history. Perhaps our intuitions concerning water are really dictated by Lavoisier's discovery of its chemical composition. We need to introduce a more dramatic variation.

So, we arrive at Version Two: this time by introducing history of chemistry. Stages One and Two are re-done: Imagine $Oscar_1$ and $Oscar_2$ living in 1750. Did each understand the term 'water' differently from the other? Again, the intuition seems to suggest that they did: the first refer to what is in fact H_2O, the second to what is in fact XYZ. So, $Oscar_1$ and $Oscar_2$ understood the term 'water' differently in 1750 although they were in the same psychological state, and although, given the state of science at the time, it would have taken their scientific communities about fifty years to discover that they understood the term 'water' differently. We perform some intuitive induction and end up with the general judgment, valid also for gold, and the like.

But what about the kinds of stuff that are both present around, perceptually indistinguishable, and strongly associated with some familiar product; or at least one of them is. This suggests further variation.

Version Three: "we will now suppose that molybdenum is as common on Twin Earth as aluminum is on Earth, and that aluminum is as rare on Twin Earth as molybdenum is on Earth. In particular, we shall assume that 'aluminum' pots and pans are made of molybdenum on Twin Earth. Finally, we shall assume that the words 'aluminum' and 'molybdenum' are switched on Twin Earth: 'aluminum' is the name of molybdenum and 'molybdenum' is the name of aluminum." 225.

Version Four: the final question concerns non-exotic, not rare specimens, and moves from science fiction to ordinary examples and from mass terms to kind-terms in botanic and the like; this is the elm-beech example (Putnam 1975, 226).

We arrive now to the general conclusion: meanings can vary whith psychological states remaining constant explanation: meaning depends on external causal connections.

> We claim that it is possible for two speakers to be in exactly the same psychological state (in the narrow sense), even though the extension of the term A in the idiolect of the one is different from the extension of the term A in the idiolect of the other. (Putnam 1975, 222)

How is this possible? We need explanation, and it will appeal to causal theory and division of linguistic labor. Further steps lead to the final reflective equilibrium, strongly revisionary of traditional Fregean semantics (for a historical overview see Floyd 2005). Here are then the stages.

Stages	Events
Stage 0—design	
Stage 1 the question	The scenario presented and the question: To what does the Twin Earth speaker refer with her word "water"?
Stage 2 understanding	The reader understand the scenario and the question
Stage 3 tentative conscious production: tries to imagine the arrangement and does it to one's satisfaction	Imagining the other planet
Stage 4 possible non-conscious elaboration	Possible problems: what is my body made of there? Is such a scenario possible at all?
Stage 5 Intuition: immediate spontaneous answer	*The Twin Earth speaker refers to XYZ, not to water*
Stage 6 Intuitive induction: varying and generalizing	Variations: First, other materials Next: History of chemistry, Aluminum-molybdenum, Elm-beech
Stage 7 general belief 7a explanation	Extensions can differ, although the speakers are in the same psychological state "Meanings are not in the head", causal theory, division of linguistic labor
Stage 8 search for reflective equilibrium	Discussion of description theory in semantics, criticism of it
Stage 9 reflective equilibrium achieved	Keep the general belief and the explanation, and restructure semantics

This concludes the short reminder of the original TE(s). Very soon after the original presentation, several philosophers, and Tyler Burge in particular in 1979 paper "Individualism and the mental", famously noted that the story about linguistic meaning can be extended to mental content as well. Burge developed it into an externalist theory of mental content. Again, we have a variation, but rather dramatic one, primarily at the level of immediate conclusions from Twin Earth and similar TEs, and perhaps even deeper, at the level of initial question. The reader is presented with variations $Oscar_1$ and $Oscar_2$ scenario, and now, in the 'laboratory of her mind she has to test a bit different assumption: is the content of $Oscar_1$'s mental state when thinking of the relevant metal the same as $Oscar_2$'s or not? Burge famously answered in the negative. Here is Putnam's comment on Burge from Burge 1996: "…I have come to believe that he is right" and "…I agree with his paper" (Goldberg 1996, p. xxi).

Once the new question was around, a new round of reactions followed, represented by ten famous paper reproduced in (Goldberg 1996) mostly from the eighties, when the action was at its peak. Searle (1983) insisted that the difference between contents is internal, due to the nature of intentionality, and reflexive functioning of the indexical(s) allegedly involved ("this metal"). Fodor (1991) proposed two kinds of content, and argued for the primacy of the internal, "narrow", one. Dennett came up with the idea of a "notional content" (see his contribution in Goldberg 1996) and Jackson (in Goldberg 1996) proposed a stable, compromise solution, that later developed into "two-dimensionalism". The difference in the question asked, or at least the difference in the weight assigned to the mental in contrast to the linguistic, resulted in differences in intuitions, then in generalizations arrived at, and then projected deep into the area of explanation and reflective equilibrium (see Goldberg 1996) for a nice collection of historically crucial papers). So, hopefully, the attention to stages might help us understand the further history of the Twin Earth TE.

But why care about such matters? Why go historical in the meta-philosophy of TEs, rather than stay with the usual style of topical discussion? We shall say more about the process after we take a look at the recent criticism of Putnam and of methodology of analytic philosophy in general.

5.3 Empty Thoughts: Peter Unger's Critism of Analytic Methodology

Let me start by quoting Peter Unger's interview in which he nicely summarizes his view of contemporary philosophy:

> Philosophers easily get the idea that somehow or other, just by considering things about the world that they already know, they can write up deep stories which are true, or pretty nearly true, about how it is with the world. By that I especially mean the world of things that includes themselves, and everything that's spatio-temporally related to them, or anything that has a causal effect on anything else, and so on. They think they can tell a deep story about how it

> is that all of this stuff really hangs together, that's much deeper, more enlightening and more comprehensive than anything that any scientist can do.
>
> And so philosophers proceed to write up these stories, and they're under the impression that they're saying something new and interesting about how it is about the world, when in fact this is all an illusion. To say new and interesting things about the world—and that's very hard, things of any generality I mean, or even anything interesting—you really have to engage with a lot of science. And very few philosophers do any of that, at least in any relevant way.(Boey and Unger 2014)

All this, Unger thinks, applies to Putnam, to whom several sections, and the whole chapter three of his book *Empty Ideas: A Critique of Analytic Philosophy*, titled Earth, Twin Earth and History, are dedicated.

His first criticism is simple: there is too much material, just spates of articles l!!! He said he entered "twin earth", coming up with 1,941 items, comprising 1,571 articles and 261 reviews, and the remainder in other categories. Is it good or bad? Depends on the quality of the articles. And here comes his second criticism: "…a fair lot of much of this remarkably empty literature":

> Philosophers easily get the idea that somehow or other, just by considering things about the world that they already know, they can write up deep stories which are true, or pretty nearly true, about how it is with the world. ..They think they can tell a deep story about how it is that all of this stuff really hangs together, that's much deeper, more enlightening and more comprehensive than anything that any scientist can do.

And Unger comes, surprisingly, with an alternative proposal, a new Twin Earth thought experiment. Here is just the barest sketch. He invites us to imagine someone like Lavoisier, "on the verge of discovering that earthly water is composed of hydrogen and oxygen." at the same time, the twin Lavoisier is on the verge of discovering the chemical composition of twin water. "in all their real mental powers, our two Joseph Antoines will be precisely similar, however we sensibly construe "mental powers". not only as regards what behavior each is apt to produce (in any encountered environment) but also as concerns what experiencing each is apt to enjoy, each of our two chemists is, at our start, precisely similar to the other" (Unger 2014:64). "…now, suppose each … of them "to switch places with the other, almost instantaneously. with this switch, twinchemist will be here on earth, in 1750". Of course, each proceeds with his discovery, only that it is now the twinchemist who discovers the composition of water. So, what is the big deal, Unger asks. "The capacity to "think of water" seems not to be needed for the job." Having beliefs about water appears irrelevant to being able to discover water's composition. Our belief-states have full general propensities, they react in the same way to water and twin water, Unger claims. Therefore, Putnam's line is irrelevant!!! Unger than offers a very complicated and sophisticated TE with elementary particles, to introduce what he sees as the relevant alternative scenario. We shall not reproduce it here.

Let us now pass to a brief defense of Twin Earth. First, Unger's own methodology suggests the indispensability of TEs: even if you see analytic philosophy as a bunch of empty ideas, you will have to propose TEs of your own. This is a deep inconsistency in what Unger is doing. Alternatively, if he had looked at alternatives outside of analytic philosophy and had turned to recent continental authors, he would, with his criteria of clarity, probably see it as passing from contradiction to contradiction; not a very attractive option.

Unger sees the biggest part of analytic philosophy as being empty of concrete content. and proposes an alternative relying on extremely sophisticated scientific knowledge as the background for analytic philosopher. Emptiness is for him tied to the endless production of tons of empty material. So, let me briefly address his objection, and then concentrate upon the first one.

The sources of potential philosophical knowledge are not mysterious as Unger sees them. Remember: "Philosophers easily get the idea that somehow or other, just by considering things about the world that they already know, they can write up deep stories which are true, or pretty nearly true, about how it is with the world." This simply doesn't fit the Twin Earth TE. The semantic part is about us. We might be unaware of the structure of our semantic habits—the reflection on cases can make us aware of it. As far as the non-human world is concerned, the info in the theory of Twin Earth TE comes from science, chemistry etc. and is then brought together with our understanding of meanings.

This is, it seems to me, quite a general pattern with famous TEs in descriptive-explanatory theoretical philosophy. For instance, Jackson's Black and White Mary makes sense only on the background of quite developed neurological picture of human mind, and questions one central ambition of it. The same with Putnam's Brain In a Vet. One element that made it interesting was his replacing of the Cartesian thinker, ego, or whatever, with brain, and his connecting this to the up-to-date debates concerning Turing machines and Skolem-Lowenheim theorem (in chapter one of *Reason, Truth and History*). I think Tim Williamson is right about the abuse of the epithet "empty" by Unger:

> Unger's use of the term "empty" is just an advertising trick. It's like a competitor who defines "empty" as "containing nothing but brand X fruit juice" and then puts up posters warning that cartons of brand X fruit juice are empty. To read Empty Ideas, one must get through the equivalent of numerous elaborate descriptions of cartons of brand X fruit juice of various types, each concluding that the carton was empty, and for contrast some elaborate descriptions of cartons of brand Y fruit juice of various other types, each concluding that the carton was full. The reader's task is made no easier by Unger's loquacious, attention-seeking prose. (2015: 22–23)

We should then turn to meta-philosophy and history of philosophy. What about Unger's first criticism? Does the spate of articles make any sense? Obviously it does for Unger, for he tries to add to it another version of Twin Earth TE. At the same time, it goes on his nerves. Here is a therapy: let us try to understand where the spates are coming from.

5.4 Stages and Traditions—From the Synchronic to the Diachronic

Unger is obviously right about TEs generating spates of articles, full of comments, criticisms and alternative proposals. But Descartes has generated more spates than any analytic philosopher. And the has provided the edition of his Meditation accompanied by criticism and his answers; leading intellectuals of his time, from physicist Mersenne, through philosopher and theologian Arnauld, to Thomas Hobbes have voiced their criticism, creating the first 'spate' announcing many others to come. Where do these spates come from? How does a spate begin? In order to answer the questions we have to dig deeper into the matter, and ultimately, we need some meta-philosophy explaining the history of TEs and in particular the recent history of analytic philosophy. First, we can note that spates come in trails. Unger himself notes the following:

> Just a few years later and largely thanks to their contemplating Putnam's Twin Earth scenarios, several other philosophers did endorse this further idea, concerning who thinks what, the two most timely being, perhaps, Burge and McGinn (2014: 77, 60)

But the spate of articles that come out of Putnam's and Burge's efforts constitutes a trail, the externalist one. We can even call it a tradition. There is the contrary trail-tradition, represented by more internalistically oriented authors, from Searle to Jackson.

The first thing to note is that the trails start as reactions to particular features of the original Twin Earth TE, and the second, and a surprising one, that reactions target particular stages of the TE, particular sore-fruitful spots which one can use as starting points from one's own proposals. Some authors, perhaps most famously Burge, widen the range of examples (from water to arthritis, and from true beliefs to systematically false ones), thus targeting the initial design of the TE and changing a bit the question asked, others stay with original example, but, at the stage of intuition-generation suggest a different intuition: some content of Twin Earthling's thought is identical to the content of Earthling's thought. The new intuition demands new explanation, and commands a different process of intuitive induction.

In other words, *there is a connection between the synchronic structure of stages and the diachronic process of generation of trails, even traditions.* Let us take a closer look at the externalist and internalist trails.

Start with Tyler Burge. Here is his question, stage 1: Does the ignorant person who believes that he has arthritis in his thighs refer to arthritis? Yes, the intuition suggests (at stages 4–5). The same for sofa and contract. We can see his initial proposal as a variation of the crucial example, replacing water with items like sofa, arthritis or contract. This changes the TE already in its initial stages. Now, intuitive induction, our stage 6: the contents of the thought of ignorant persons all depend on the public meaning of the terms involved. The very content of thought is determined externally; a strong externalism about content, not just about linguistic meaning, follows. Of course, a lot of work is to be done at the next stages to relate the view to other accepted ideas in philosophy of mind. Burge and his followers will be adamant about the externalist intuition, its externalist explanation, and the general externalist stance; all else can be changed, but not these results. Some kind of reflective equilibrium follows (of course, it can be challenged by the internalist, who might point out that on this account we don't know contents of our thoughts, as Boghossian famously did-see References).

Incredibly enough, the trail generated by the externalist proposal will be crossing another trail, the Demon TE that comes from a Cartesian tradition, and the fans of the trail will try to subvert the results from the Cartesian one. Consider a brain-in-a-vat: does it have false thoughts about the external world, a Cartesian thinker will ask. Well, does it have thoughts about the external world at all? The externalist line suggests a negative answer. But, if it has no thoughts about the external world, it does not have any false thoughts about the external world; skepticism is inconsistent. No wonder, there is a recent book bringing together the papers from the crossing, Goldberg's (2016) collection *The Brain in a Vat*. A discussion of this crossing of the two traditions would demand a separate chapter. However, mere pointing out the phenomenon of crossing is all we can do here.

To conclude, the history of semantic externalism can be reconstructed to a large extent as history of the Twin Earth TE(s) debate. Attention to the internal structure of the TE(s), especially to the sore-fruitful spots at each stage of the TE that prompt questioning and the debate make see the discussion and the 'spate of articles' produced as rational response to the crucial issue involved. The elementary reconstruction of the TE(s) in terms of stages, and of problems connected with the performance of the experiment combine with diachronic rational reconstruction of the history of externalism as a whole and make us understand the important trail of contemporary analytic semantics and theory of mental content in rational and philosophically relevant terms.

Let me now pass to the opposite, internalist trail, extremely popular, with a lot of followers. First, a moderate proposal. Searle insisted that the difference between contents is internal, due to the nature of intentionality, and reflexive functioning of the indexical(s) allegedly involved ("this metal"). *Intentionality,* pp. 206 ff. This attacks the explanation. Note that there is no variation in the TE itself, but considerations of causal relation between thoughts, objects and actions.

Next, consider a more radical proposal, involving a deeper analysis of earthling vs. Twin earthling thoughts. There is a kind of mental content, narrow content, that is the same across Earth-Twin Earth contrast. Just consider causal powers of the thoughts: "being a water thinker is the same causal power as being a twater thinker, only instantiated in a person with a different causal history." (Fodor 1991, 5–26, reprinted in Pessin and Goldberg (eds.) 1996, 275 ff). His further conclusion is that the narrow content is the right content for psychology (and psychological generalizations), the view known as methodological solipsism The final reflective equilibrium reached is completely opposite from Burge's and later Putnam's, and it has had a lot of followers.

So, we may look at the longer term history of different stages within the history of a TE or its variants. It starts with initial experiment, including variation, general belief-conclusion, explanation, and early reflective equilibrium, and can for example proceed to a new question. Is the linguistic meaning determined by external factors is supplemented with a more dramatic question: isn't the mental state itself then determined by external factors? There will be a new example (arthritis in the thigh), or a new reading of the old one (or one can accept two kinds of content, like Fodor, and Jackson). The new process of 'intuitive induction' will lead to a new conclusion, say that one extended to mental content that now varies with the surrounding. Then we need a new explanation: causal theory is widened to the mental content, and the division of linguistic labor has internal psychological consequences. This is a new reflective equilibrium, different from the earlier one(s). With some luck a new trail is created, with can become a real new tradition. And this can be repeated many times, passing, for instance to analysis of ethical discourse, and proposing Moral Twin Earth TE (see Horgan and Timmons 1992). But, there is space for other kinds of variation, for instance, including information about scientific treatment of kinds (chemistry, biology, etc.), what Unger would describe as Substantial Scientiphicalism. Or, one can proceed to psychological questioning as has been done in experimental philosophy (see Haukioja 2015 for some discussion of Putnam).

What is crucial of us is the link between short-term and long-term pattern of (thought-) experimental reasoning. This has not been noticed in the literature. Let me summarize what has been said in this section in the following table:

	Variation and new trails	Examples
Stage 0 design	Accommodating data from stages 6, 7, 8 and 9	Redesigning the TE • Tyler Burge: externalist proposal • Jackson: varying which world is actual • Fodor: methodological solipsism
Stage 1 the question	New question: reformulation or complete transformation	Widening: • From language to the mind or • From mind to ethics: moral Twin Earth
Stage 2 understanding:	(adding explanatory material)	
Stage 3 tentative conscious production: tries to imagine the arrangement and does it to one's satisfaction. Stage 4 possible non-conscious elaboration Stage 5 Intuition: immediate spontaneous answer	Questioning intuition, suggesting a different judgment	Jackson: a very different intuition XYZ is water in one sense, is not water in another Fodor: We and Twin Earthlings share one content (narrow), but not the other (wide)

(continued)

(continued)

	Variation and new trails	Examples
Stage 6 Intuitive induction: varying and generalizing	• Conservative: Widening the range • Revolutionary: Problematizing the original examples • ????	Jackson and Fodor both widen the range enormously, so as to cover all Putnam's cases (and perhaps more)
Stage 7 general belief 7a explanation	A new general belief G: • Conservative extension • Revision • ??? A new explanation	A new general belief G*: Jackson: Two meanings Fodor: 2 kinds of content methodological solipsism
Stage 8 search for reflective equilibrium	• Traditionalist: revise G and keep the inherited beliefs • revolutionary: keep G and change the inherited beliefs	
Stage 9 reflective equilibrium achieved	The new theory: • conservative • revolutionary	

But why do philosophers go through all these efforts. One answer is that TEs are indispensable. Philosophy does not use laboratory to test its theories; the only experiments available here are those in thought. TEs play in philosophy the crucial role that laboratory experiments play in science. Philosophers are vitally interested in *connections between our spontaneous understanding of important items, like meaning and content of our thoughts, and the results of science.* In order to answer the question about the relation between, say, cognitive science-cum-neurology and our feeling of having contentful thought, we need to confront the two, and we cannot do it within science alone. We need the bridge, and a TE is a perfect candidate. TEs are our laboratory, and philosophers return to their experiments, as scientists do to theirs.

We now understand why one should go historical in the meta-philosophy of TEs, rather than stay exclusively with the usual style of topical discussion. The spates of articles, lamented by Unger, are trails, promising to become traditions, and mark the present-day analytic philosophy, as their ancestors marked the early twentieth century efforts.

5.5 Trails-Traditions

It is time to generalize, and draw the morals from our discussion of Putnam's TEs. We obviously need a historically informed meta-philosophy of analytic efforts, in this case ones connected to thought-experimenting, and combining the awareness of the inner structure of a TE with insights into historical, diachronic process of generation of trails leading from various components of a TE to further and further developments. Some famous examples, much older than the analytic tradition, like Plato's TE of ideal state, and Descartes's Evil Demon TE present the same trail-forming tendency.

Here are the elements we were able to identify. First, the indispensability of thought-experimenting. There is often no other way to address central philosophical questions. So, there is no wonder that challenges TEs are revisited time and again.

Second, performing a TE involves asking a question and then letting the subject decide; like in laboratory TEs the designer is not the absolute master. There are performance challenges, issuing in hot spots: the reader responds differently from the expectations of the designer. So, the designer has to go back to the drawing board, and vary the initial proposed scenario. Or, a colleague with different background opinions might find the intended answer incredible. Then she has to propose an alternative scenario that will vindicate her intuitions. This produces endless variation in scenarios and questions and production of new intuitions and immediate generalizations.

Next, the history sets in, with long term development of these generalizations, attempts to explain them, and new candidates for reflective equilibrium, from very conservative to very radical ones. Trails are being born, that, if there is luck enough, turn into solid tradition.

Note that laboratory experiments are discussed and varied in the same way as TEs. Take Edison's discovery that direct current can do marvelous things. But then comes Tesla: "Yes, boss, but alternating current can do many more, and much more marvelous ones." What followed was "War of Currents" in the 1880s over whose electrical system would power the world—Tesla's alternating-current (AC) system or Edison's rival direct-current (DC) electric power. So with laboratory experiments we have the same pattern: first, the initial experiment, followed by small, routine variations, then, in case of really important ones, a spectacular variation. Scientists discuss the prospects of both, and propose further variations, if needed. Remember the quote from F. James who stresses the variability of experiments, and compare his diagnosis to what happened with Twin Earth TE.

Back to the diachronic developments in a long-term life of a TE. Let me make a first step towards generalizing the morals of it and offer a new proposal for the understanding of the history of debates around important TEs that make up a significant part of analytic philosophy. Let me again mention two to three famous TEs: Evil demon-Brain in a vat TE and the Social contract. These have engendered long term traditions, each lasting around three hundred years, if we count modern philosophy only, and set aside the Greeks and Saint Augustine. If we could bring together short-term and long-term stages, this would make the historical process more intelligible.

For instance, take the relatively recent variations of the two mentioned thought-experimental oldies. The Evil Demon has been innovatively transformed into (or replaced with) the Brain-in-a vat (see Goldberg 2016), and Social contract has generated the Original position proposal (see Hinton 2015) and its relatives, due to Scanlon and Habermas.

The patterns seems recognizable. The tradition starts with an initial TE, and proceeds through three possible kinds of reactions. First, a new variant of the TE or a new, but related TE is proposed. In the Social Contract tradition, one can change the characterization of the parties involved (males only, females as well, what about mentally challenged persons, and so on). The dramatic example of the latter is Brain-in-a-vat: replace the thinking person from the original Cartesian scenario with a brain wired to a control panel. The same with science: in the next chapter we shall sketch the famous Einstein–Podolsky–Rosen TE that has generated a long and fruitful trail, reaching to our days.

Or, there can be a non-thought experimental reaction to the initial TE: think of the numerous objections to Descartes *First Meditation*, collected in *Meditations with Objections and replies*, or of the "negative program" within experimental philosophy, to be discussed in our next chapter, or, concerning Original position and its consequences, the criticisms of B. Barry).

Finally, we can have both, criticisms plus new TE, the way we just saw in Unger's reaction to Putnam. Once we have the big picture, we understand the strange reaction from Unger: on the one hand criticizing the spate of papers discussing Twin Earth TE as empty, meaningless, and over-complicated, and on the other, adding to the spate, by proposing further, related, but much more complicated TEs. On the one hand, a philosopher or a scientist is tempted to a negative reaction when one notices the sheer length of a famous TE trail. On the other, the only legitimate way to criticize is to add to the trail!

The history is in some respect analogous to short term reflection and discussion of such a TE, say in a seminar or a conference. Stages of understanding and debating normally structure the short term life of the TE. Longer lasting attempts (with variations, often fundamental) are part of the further, historical story. So, we can integrate the view of history of (analytic) philosophy with meta-philosophy of TEs, and then, we shall find Unger-like disappointments and criticisms less persuasive, and less worrisome. Twin water has become part of the philosophical ecology, and has a right to stay with us. It is part of our life as philosophers. In the case of science, understanding the history of science, we need the understanding of the longer history, of the laboratory experimental tradition tied to some important initial experiment. The same holds for TEs in science: we need the understanding of the longer history of each given TE in the history of physics. The debate between Einstein and the proponents of quantum physics has left us a rich heritage of such discussions of TEs. We want to know what relation holds between such historical changes both in philosophy and in physics, and the usual micro-variations of a given TE. Once we have a worked out answer we shall be able to integrate the meta-philosophy of TEs with information from history of philosophy and theory of science. Here, we have just made the first steps.

References

Boey G, Unger P (2014) Philosophy is a bunch of empty ideas: interview with Peter Unger, Monday, June 16. http://www.3quarksdaily.com/3quarksdaily/2014/06/philosophy-is-a-bunch-of-empty-ideas-interview-with-peter-unger.html

Borges R, Almeida CD, Klein PD (eds) (2017) Explaining knowledge: new essays on the gettier problem. Oxford University Press, Oxford

Burge T (1996) Other bodies. Reprinted In: Pessin A, Goldberg S, pp 142–160

Floyd J (2005) Putnam's "The meaning of 'meaning': Externalism in historical context. In: Y Ben-Menahem (ed) Hilary Putnam. Cambridge University Press, Cambridge, pp 17–52

Fodor JA (1991) A modal argument for narrow content. J Philos 88(1):5–26

Goldberg SC ed. (2016) The brain in a vat. Cambridge University Press, Cambridge

Gooding D (1989) History in the laboratory: can we tell what really went on? In: James F (ed) The Development of the Laboratory London, Macmillan

Hacking I (1992) Do thought experiments have a life of their own? Comments on James Brown, Nancy Nersessian and David Gooding. In: PSA: Proceedings of the biennial meeting of the philosophy of science association, vol 1992, Volume Two: Symposia and Invited Papers Chicago: The University of Chicago, pp 302–308

Putnam H (1975) The meaning of 'meaning'. In: Mind, language and reality. Cambridge University Press, pp 215–270

Stuart M, Fehige Y, Brown JR (eds) Routledge companion to thought experiments. Routledge, London, pp 526–545

Unger P (2014) Empty ideas: a critique of analytic philosophy. Oxford University Press, New York

Chapter 6
The Challenge of Experimental Philosophy

6.1 X-phi-the Negative Program

6.1.1 Introduction

We have briefly mentioned some criticisms of TEs, and tried to defend TEs from them. In this chapter we focus on a recent critical movement, the so called "experimental philosophy", X-phi for short that has been in the center of discussion of TEs in the last two decades. The leading idea has been to test naïve intuitions from TEs and apply the usual psychological methods, above all statistical techniques in doing it.

X-phi has raised questions about TEs and intuitions, in connection with the issue of rationality: how rational are people's intuitions? A popular distinction here is the one between the positive program, enacted by authors like Haidt (2001), Prinz (2007).[1] and the negative one, incarnated by Stich, Knobe, Weinberg, Nichols, Machery and Mallon (see References). Here we are more interested in later. In the initial, more radical phase, the negative X-phi played with a drawing representing the burning armchair; this was the way to symbolize the X-phi attack on philosopher's armchair. Ron Mallon once wrote that "the negative program of the experimental philosophy is just attacking the idea that our intuitive judgments of the world are a way getting at facts that we ought to trust." (on the old X-phi internet site, not reachable any more). The huge majority of the TEs tested has been philosophical, some little work has been done in neighboring areas, like semantics, on the frontier with linguistics and on some topics closed to cognitive science; TEs in science have almost not been addressed at all, but they might get more attention in the near future.[2] So, in this chapter we shall mostly stay with philosophy, and mostly with TEs which we have already summarized in previous chapters, above all Gettier and Truetemp; we shall briefly turn to TEs in science in the last sub-section.

---

[1] See a fine overview in Mortensen and Nagel (2016:53–71).

[2] For a recent overview see the section "Philosophy of Science" in Sytsma and Buckwalter (eds.) (2016).

N. Miscevic, *Thought Experiments*,
https://doi.org/10.1007/978-3-030-81082-5_6

The negative experimentalist started questioning TEs using examples of irrelevant influences[3] The authors targeted were, on the one hand, and most importantly, the traditional TE fans, and on the other positively oriented experimentalists.[4]

Two kinds of problems have been in the foreground: first, the alleged: irrelevant factors influencing people's thought-experimenting and various doubts about competence of thought-experimenters, first the ordinary ones, and later of the experts as well.[5] Interestingly, the same broad kinds of irrationality have been detected by cognitive scientists working on human inference in general, for instance by authors, like Daniel Kahneman, who claim that human cognitive system contains two inferential sub-systems, one slow and careful, but costly, and the other cheap and quick, but unreliable. Such moderate 2-system theory might be correct. Others, like Sperber talk about massive modularity, resulting in very similar phenomena. If these approaches are right, the negative X-phi discoveries are just part of an overall phenomenon of human irrationality, which then appears less specific and less threatening for TEs as alleged particular area. If we are as stupid as negative X-phi suggests, why did our cognitive ability finally work in science and math? Why does it work for language, and guide linguists in their collection of data? We shall be appealing briefly to the cognitive theories, to blunt a bit the sharpness of X-phi critique.

Here is then the preview of the chapter. In the first part, we shall present some pioneering results of early X-phi program, and very briefly point to subsequent developments. In the second part, we look at the medicine for the problems diagnosed by X-philosophers. We very briefly look at two extant proposals, and add our own as the third kind of medicine, connecting it with the mental modelling and stage-distinguishing approach we sketched in the previous chapter.

6.1.2 The Challenge

How is X-phi questioning TEs? Here, we can offer only a very brief sketch. It all started at the turn of the millennium, with investigations of the following issue: are people in performing TEs being guided by relevant factors, or are they typically victims of completely irrelevant influences. X-phi was born from the result that suggested endorsing the later alternative. We shall here mostly mention these results, and just briefly point further to more recent spate of results, that would require a separate book to be presented and discussed properly.

So, let me remind you of the most spectacular pioneering examples of the irrelevant influences on all sorts of thought-experimenting, the ones due to affective content and to affective context.[6]

[3] For a useful overview of huge literature on the topic see Buckwalter (2016) (eds.).

[4] For a relatively recent list of problems by Alexander et al. (2010).

[5] The pioneering works are Weinberg et al. (2001), Nichols et al. (2003a, b), Machery et al. (2004).

[6] The main early sources are Haidt (2001), Nichols (2004), Nichols and Knobe (2007), Prinz 2007 (recently summarized in Stich and Tobia (2018:374)).

The first are examples of affective surroundings: intuitions are sensitive to *affective context*, such as seeing and smelling disgusting stimuli like greasy pieces of old pizza (Schnall et al. 2008); the subject's moral judgment is *more severe* in such circumstances than in the neutral ones!

The next group of irrelevant factors concerns affective content. The most famous work that has helped to started the X-phi program, Nichols's and Knobe's (2007) paper, stresses the dramatic role such content is playing in naïve ascription of responsibility. Here is the famous vignette that was offered to the subjects:

Affective Content.

Imagine a fully deterministic universe. The agent there does

(A) a morally neutral deed, (B) a murder: wife and kids by setting fire, in order to continue his affair with his secretary.

Is the agent there responsible for his deeds? (A) no, (B) yes (72%).

Subjects' intuitions about the relationship between causal determinism and moral responsibility can depend on the presence or absence of affective content in the narrative of the thought-experiment (Nichols and Knobe 2007). Such data indicate that intuitions about the relationship between causal determinism and moral responsibility are unstable.

Ascription of Intentional Doing Cond. 1 A Program that Harms:

- **Shall We Start a Program that Harms the Environment?**
- The chairman of the board answered, "I don't care at all about harming the environment. I just want to make as much profit as I can. Let's start the new program."
- They started the new program. Sure enough, the environment was harmed.
- Now ask yourself: Did the chairman of the board intentionally harm the environment?

Condition B: A Program that Helps

- Vice-president said, "We are thinking of starting a new program. It will help us increase profits, and it will also help the environment." The chairman of the board answered, "I don't care at all about helping the environment. I just want to make as much profit as I can. Let's start the new program."
- They started the new program. Sure enough, the environment was helped. Knobe (2003). "Intentional Action and Side Effects in Ordinary Language." Analysis, 63, 190–193.
- The results were clear and compelling: 82% of subjects who received the story about environmental harm said that the chairman harmed the environment intentionally, whereas only 23% of subjects who received the story about environmental help said that the chairman helped the environment intentionally.

In short, according to Alexander and Weinberg, these data show that intuitions are "sensitive to factors irrelevant to the content of the thought-experiments themselves"

(2007:61). In other words, people's intuitions are influenced by "irrelevant" factors. Sensitivity to irrelevant factors, argue Alexander and Weinberg, "impugns the status of intuitions as evidentiary" (2007:66).

The pioneering paper mentioned has generated a spate of literature, both experimental and wider methodological. The researchers have naturally turned to related issues of responsibility, and to the question whether free choice and responsibility are compatible with determinism. Here we cannot go into this rich and impressive new research tradition and will limit ourselves to suggesting consulting a very useful overview by Chan et al. (2016), who conclude that "/t/he experimental philosophical work on free will thus does not support the view that there is a unified, coherent ordinary concept of free will/moral responsibility that can be adequately captured by either compatibilism or incompatibilism; it appears that the way people think about free will and moral responsibility contains both compatibilist and incompatibilist elements" (2016:68).[7]

The next source of irrational reactions concerns the order of presentation. You tell the story from the TE you are interested in, and then you accompany it with comparable stories, differing from it. The order of presentation will influence subject's judgments in the case of the TE. The most famous case; it concerns the Truetemp TE that we recounted in chapter Two. Now, the experimenters note that subjects' intuitions about Truetemp-like cases were apparently influenced by the presentation of contrasting cases, in particular by the order of presentation of the various examples. The two competing examples were mutually contrasting: the first, a clear example of a lack of knowledge, and the second of knowledge. The first featured Dave, a man who believes he can predict when a coin will land heads. Dave's predictions of the outcome of coinflips are no better than chance, but in the case at hand he happens to get it right (Question: Does Dave know?). The contrasting clear case of knowledge, was the one involving a knowledgeable chemist named Karen who gained the relevant information by reading a leading scientific journal. It was expected that a very clear case of knowledge would make the Truetemp Case seem less plausibly a case of knowledge by comparison. The Chemist Case is a clear case of non-controversial knowledge of a trained expert. The authors recount the result:

> Our results show that compared to subjects who receive the Truetemp Case first, subjects first presented with a clear case of knowledge are less willing to attribute knowledge in the Truetemp Case, and subjects first presented with a clear case of non-knowledge are more willing to attribute knowledge in the Truetemp Case. We contend that this instability undermines the supposed evidential status of these intuitions, such that philosophers who deal in intuitions can no longer rest comfortably in their armchairs. (Swain et al. 2008), reiterated in Stich and Tobia (2018:373),

Interestingly, in other, more general experiments on rationality (Kahneman 2014, employing examples from business decisions) the same effect of sensitivity to the order of presentation was found. So, it is not peculiar to TEs and to philosophy.

[7] For more see Cova (2016) and Björnsson and Pereboom (2016).

We now pass to the examples suggesting that the cultural and social background of the subjects influences their judgments.[8] A study by Machery et al. (2004) concerned intuitions about reference. Saul Kripke famously argued that reference of names does not follow associated descriptions. He starts by reminding us that the only thing most people know about Gödel is that he proved the incompleteness of arithmetic. Next, he invites us to suppose he hadn't proved it, but had in fact stolen it from his friend Schmidt. Now, would the name "Gödel" then refer to Gödel or to Schmidt? There is an agreement in universities in US and Europe that it would still refer to Gödel. Machery and his collaborators found that subjects with Western ethnic origins were more likely than those with East Asian ethnic origins to give causal-historical responses to the queries.

Other studies showed variation in intuitions across (American) subjects with different ethnic origins. Weinberg et al. (2001) gave their subjects a Gettier-like example featuring a protagonist named "Bob," and asked them whether Bob "really knows" a specified proposition or "only believes" it. A majority (74%) of subjects with Western ethnic origins responded that Bob only believes it, while a majority of subjects with East Asian (56%) and Indian (61%) ethnic origins responded that Bob knows.[9]

Remember Gettier's story of Smith, Jones and Brown; Weinberg, Nichols, and Stich have in their (2001:29) retold a version of it for the experimental subject:

> Bob has a friend, Jill, who has driven a Buick for many years. Bob therefore thinks that Jill drives an American car. He is not aware, however, that her Buick has recently been stolen, and he is also not aware that Jill has replaced it with a Pontiac, which is a different kind of American car. Does Bob really know that Jill drives an American car, or does he only believe it?

And they offer two answers concerning Bob:
Really Knows Only Believes.

The commentators have discussed such examples of intercultural differences in judgment about knowledge- is the difference conceptual or perhaps rather factual- e.g. in epistemology the assumptions about reliability of one's memory, or in the Gettier 10 cents story of standard authoritative testimony. If the first, the difference is principled and we, thought-experimenting see are in trouble, if the second, not so much. Results vary. The early investigations came up with dramatic variations, the later ones with much smaller. See Nichols et al. (2003/2012) and Machery (2014), and Machery et al (2016). The typical recent characterization is the one by Stephen Stich Masaharu Mizumoto and Eric McCready, from their "Introduction" to (2018), 1–11.

> More than half the authors are philosophers, but the others are linguists. They discuss various issues surrounding what is called the "universality thesis" in this book's manifesto—the thesis

---

[8] For recent short summary see Stich and Tobia (2018:374).

[9] A different study by the same authors (Nichols et al. 2003a, b) reported that the number of philosophy courses subjects had taken was a factor in influencing their responses. Subjects who had taken a number of philosophy courses were more susceptible to skeptical arguments than those who had taken fewer.

> that, roughly speaking, properties of the English word "know" are shared by counterparts in most or all other languages. Some argue directly against this thesis, some defend it, and others discuss methodological or metaphilosophical issues surrounding the debate. P.1

We shall skip here the results of investigations concerning the way people react to questions in moral TEs.[10]

This brings us to the conclusion. We shall here stress the negative take-home messages from X-phi, which are the most challenging ones, and which have played the central role in discussions. So, we pass to the pessimism generalized, the view that intuitions are hopeless. Weinberg writes:

> "A source of evidence that is not practically infallible is hopeful to the extent that we have the capacity to detect and correct its errors. We stipulate (…) "hopeless" for devices for which our practices lack an appropriate sensitivity to their errors, and capacity for correction when such errors are found.
>
> (H): Any putative source of evidence that is hopeless ought not be trusted. (Weinberg 2007:327)

And he explains that intuitions lack the 4 sources of epistemic hopefulness:

- no external corroboration!
- no internal coherence !
- no testability of margins of reliability!
- no theoretical illumination! (p. 330)

His conclusion: intuition is hopeless as source of evidence!

Eduard Machery in his recent book (Machery 2017) follows a similar line, and systematizes the weaknesses discovered by experimental research in three main groups (we shall use his systematization in the sequel.) First, the lack of reliability of cognitive processes in TEs, documented somewhat by their lacking hopefulness and calibration, but mainly by uncontrolled variation: "Judgments elicited by most of the philosophical cases that have been examined by experimental philosophers are unreliable", and he suggests the following: "Philosophers ought to suspend judgment about the situations described by philosophical cases" (2017:103).[11]

---

[10] See for instance recent surveys in Greene (2016) and Justin W. Martin and Fiery Cushman *The Adaptive Logic of Moral Luck* (2016).

[11] Some of the more recent comments by X-philosophers are less dramatic. For instance, Stephen Stich and Kevin Tobia recently offered a more modest proposal:

> The skepticism about the use of intuitions as evidence (…) should be directed primarily at intuitions that have been shown to be susceptible to irrelevant influences, and at other intuitions that are the product of psychological mechanisms that are likely to be vulnerable in similar ways. And which are these? The answer is that at this point we do not know, and we are not going to find out without a great deal more sophisticated work in psychology and neuroscience. On our view, this sort of research should be viewed as quite central to experimental philosophy. So the take-home message from this chapter is not that intuitions should not be used as evidence in philosophy. Rather, it is that experimental philosophy, broadly construed, has a crucial role to play in assessing and improving philosophical methodology (2018:379).

Here is his complete Argument from unreliability:

1. Unreliable judgments are severely deficient from an epistemic point of view.
2. Judgments elicited by most of the philosophical cases that have been examined by experimental philosophers are unreliable.
3. If the judgments elicited by most of the philosophical cases that have been examined by experimental philosophers are unreliable, then the judgments elicited by most philosophical cases are plausibly unreliable.
4. We ought to refrain from making a judgment of a particular kind K (i.e., we ought to suspend judgment of kind K) when most judgments of this kind are plausibly severely deficient from an epistemic point of view, except when this judgment is known to be an exception.
5. Hence, except when a philosophical case is known to elicit a reliable judgment, philosophers ought to suspend judgment about the situations described by philosophical cases (p. 103).

The second group concerns what he calls "dogmatism". "Most of the philosophical cases examined by experimental philosophers elicit disagreement", he writes (2017:128), and in such cases it is dogmatic not to suspend the judgment about the matter.[12]

The third group concerns variations across groups, cultures, nations and the like; he speaks of Parochialism: When people respond differently to the philosophical cases examined by experimental philosophers, it is because they refer to different (epistemic, moral, etc.) properties (2017:140); he means each group referring to a different property (knowlede-for-Asians vs. Knowledge-for-Europeans). So, we should give up relying on Tes (on "method of cases" as he calls it.)[13]

---

[12] Here is his complete Argument from dogmatism:

1. Most of the philosophical cases examined by experimental philosophers elicit disagreement.
2. This disagreement takes place among epistemic peers.
3. If most of the philosophical cases examined by experimental philosophers elicit disagreement among peers, then most philosophical cases would plausibly elicit disagreement among peers.
4. If epistemic peers are likely to disagree about a philosophical case, they ought to suspend judgment about it.
5. Hence, except for those philosophical cases known not to elicit disagreement among peers, philosophers ought to suspend judgment about the situations described by philosophical cases (2017:128).

[13] Here is his Argument from parochialism:

1. When people respond differently to the philosophical cases examined by experimental philosophers, it is because they refer to different (epistemic, moral, etc.) properties.
2. If people refer to different (epistemic, moral, etc.) properties when they respond differently to the philosophical cases examined by experimental philosophers, they would plausibly do so in response to most philosophical cases.
3. Philosophers are not justified in believing that theorizing about the (epistemic, moral, etc.) properties they refer to will allow them to achieve their philosophical goals.
4. You ought not to decide to φ in order to ψ if you believe you are not justified to believe that φ-ing is likely to bring about ψ.

Machery then proposes to replace philosophical methodology with a psychological one. He advocates "dismissing" many questions central for philosophical tradition and contemporary philosophy-mind–body dualism vs. identity, the nature of knowledge, the nature of causation, personal identity and many others. They are, he claims beyond our epistemic reach (2017:1).But there are "important issues in the neighborhood of most of the dismissed philosophical issues" (ibid.), namely what science alone has to say about each of the topics concerned. As far as methodology goes, what is left is what he calls "The method of cases 2.0", namely "presenting cases to samples of participants and analyzing their response statistically (2017:239). The method focuses "on the psychological entities that govern spontaneous inferences, judgments, and explanations" (Ibid.). And it is open to the possibility of extreme variation across groups, populations, cultures and traditions. Sounds like the modest remnants of philosophy, left after the central questions are dismissed, is being turned to psychology.

So, authors like Weinberg and Machery, and their followers, propose replacing the usual thought-experimenting with very different alternatives. We have argued that TEs are indispensable, so, in the face of X-phi some defense of TEs is needed, and indeed, it has been offered. For example, Kirk Ludwig gave a radical characterization of X-phi that makes it only minimally relevant to philosophical investigation:

> What then is the relation of x-phi to philosophy? X-phi is philosophy from the sociological point of view. From the standpoint of the tradition in philosophy, in so far as it is to be of aid to philosophy, it is more properly thought of as a non-philosophical activity or an adjunct to philosophical activity that has relevance to philosophy in the way in which the investigation of the foibles and frailties of human reasoning has relevance to the practice of human reasoning. We should use all the tools we can assemble to advance our understanding.
> Kirk Ludwig (2018:401), "Thought Experiments and Experimental Philosophy"

But this is not enough. What should we do about the doubts that X-phi generates in relation to TEs? One should do better.

6.2 Defending TEs

We have seen that life without TEs is impossible for philosophy, unless we replace TEs with its relatives from famous fiction and art, which cannot address problems raised by X-phi.

Hopefully, our discussion in chapter Two has made it clear that although life without TEs might be possible for science, it is practically impossible for philosophy. So we need a defense of TEs against the negative program.

---

5. Hence, philosophers ought to focus on determining whether theorizing about the (epistemic, moral, etc.) properties they refer to or about the (epistemic, moral, etc.) properties others refer to will allow them to reach their philosophical goals, stopping for the time being their theorizing about the (epistemic, moral, etc.) properties they refer to.
6. Hence, for the time being, philosophers should not appeal to the method of cases (2017:140).

One possible line is suggested by defenders of the positive program. Mortensen and Nagel (2016:60 ff.) suggest that the negative program colleagues are exaggerating their results, and point to research that has resulted in much more moderate verdicts. We hope they are right, but we would like to discuss the other possibility, that people are really very weak in thought-experimenting.

So we pass to other proposed solutions. We start with what is called expertize defense, also dear to Ludwig, and to elements of it that are certainly worthy to be preserved. Than we formulate and additional strategy, connected to our mental modelling project, and the interest in the biography of TEs; we shall try to integrate into this strategy a third, pacifistic, line of answer which rests content with the possibility that main variations come from the fact that people have not one relevant concept (say of knowledge, moral permissiveness, or responsibility) but several ones.

And, as a preliminary work, we connect the issue briefly but more firmly with theories of (i-)rationality (Kahneman, Sperber), which was not done systematically (strangely). This fits with Ludwig's suggestion of the cognitive "investigation of the foibles and frailties of human reasoning" but makes it directly relevant to philosophy. Finally, in regard to the general topic of the book, our tentative new contribution to the defense of TEs, compatible with most elements of expertise defense, claims that attention to stages in the life of a typical intuition could contribute to alleviating worries, and in fact situating the TE-activity within an acceptable framework of understanding of rationality.

6.2.1 *The Expertise Defense of TEs*

Various answers have been proposed to various problems unearthed by negative programmers. But we need a systematic framework. Here we shall concentrate upon the expertise proposal, the most successful proposal until now.

It starts from the widely accepted view that experts are better than novices in thought-eperimenting: they have streamlined concepts, worked out theories and skills in philosophical reasoning. And it passes to the obvious solution: don't ask novices, ask experts. Here is Kirk Ludwig's early formulation of the proposal:

> Clearly, training, both on the part of the experimenter and on the part of the subject, is important for well-conducted thought experiments. It is important that the experimenter know what the methodology of thought experiments is and what sorts of difficulties arise in conducting them and getting useful results. It is also important that the experimenter have some knowledge of prior work on the conceptual field in which he is doing work and where the problem areas are in investigating the structure of the concepts he is concerned with. The subject of the experiment should likewise be apprised of the methodology of thought experiments so that he knows what task is being set him and what his responses are supposed to be based on./

Why is expertise important? While "/O/therwise, he (the subject-NM) may well give a response which does not answer to the purposes of the thought experiment. He may think he is being asked to say what people would say, or he may not recognize that he is not to imagine himself in the situation, or to set aside contingent beliefs about how the world is, or to look for biases which may infect his conception of the scenario. Or he may just say the first thing that he thinks without attention to trying to answer on the basis of what the concepts involved dictate" (Ibid).

In the last decade, the prominent expertise defenders have been Sorensen (2014), Jackson (2011). Devitt (2011, 2012), Williamson (2007a, b, 2011) and Kauppinen (2014). Here are two fine quotations from Williamson, that illustrate the line of defense:

> […] philosophy students have to learn how to apply general concepts to specific examples with careful attention to the relevant subtleties, just as law students have to learn how to analyze hypothetical cases. Levels of disagreement over thought experiments seem to be significantly lower among fully trained philosophers than among novices (2007, 191).
>
> […] We should not regard philosophical training as an illegitimate contamination of the data, any more than training natural scientists how to perform experiments properly is a contamination of their data. Although the philosophically innocent may be free of various forms of theoretical bias, just as the scientifically innocent are, that is not enough to confer special authority on innocent judgment, given its characteristic sloppiness' (2007, 191).[14]

Michael Devitt notes that "there is no evidence of variation in the intuitions of philosophers (2012:14), but also warns the readers that the defense is not simple nor easy (2012:19). Sorensen talks of the Resilience of Superior Performance in his "*Novice Thought Experiments*" (2014). But what is expertise: the 10,000-h rule, borrowed from cognitive research says that expertise requires 10,000 h of practice, he reminds us. And all the expertise defenders point to parallels with expertise in science.[15]

---

[14] And one more:

> […] philosophical education (…) is supposed to improve far more mundane skills, such as careful attention to details in the description of the scenario and their potential relevance to the questions at issue (Ibid.).

[15] For instance, Jackson notes the following:

> Physicists aren't being chauvinists when they tell us which ways of categorizing systems of particles are important. They are doing their job. Part of our job—the job of philosophers—is to assess which concepts in, for example, epistemology are theoretically important, to delineate the connections between various concepts, and to identify the questions worth asking. 2011.480.
>
> Here it is worth recalling the history of probabilistic reasoning. Early folk concepts were progressively refined. (…) Polls are relevant to the elucidation of one or another folk concept relating to our uses of terms like 'expect' and 'believe likely'. They are not relevant to the assessment of the edifice of probabilistic reasoning that informs current statistics. This is why the 'they are wrong' response is the correct one to those who commit the gambler's fallacy. We can think of philosophical work in epistemology as aiming to build on some of the same folk materials, with the aim of producing a finished theory (a theory which of course will incorporate probabilistic ways of thinking). Polls won't be relevant to assessing the final product (Ibid).

The problem that has been noticed in the literature is that the defense is, in a way, isolating philosophers: they don't address issues people normally think about, or if they do, their intuitions are incommensurable with ordinary ones. As Machery has recently put it, such a method "could only contribute to the analysis of the analyst and her colleagues' concepts, and it would be of no use to examine how some group of non-philosophers conceives of something (e.g., to understand how cognitive psychologists conceive of innateness). (2017:234. Indeed, since the times of Socrates, we, philosophy teachers like to think of ourselves as experts who address certain worries of ordinary people, and who share an initial data bases with them, namely the one provided by ordinary intuitions. In modern age, with thinkers like Descartes, the basis was widened to include general issues about science, and worries that might hunt the conscientious scientists. Nowadays, it is probably issues like the ones of *The Fight Against Doubt: How to Bridge the Gap Between Scientists and the Public*[16]where philosopher join hands with scientists in defending practices like vaccination, and projects like fighting the climate change.

I want to stress problems that have not been yet discussed. The defense starts from the final shape of philosophical expertise. But how does one become expert in philosophy? If our ordinary opinions are worthless, where does the "production" of good academic philosophers start?

The question is analogous to the ancient problem of becoming wise. If you asked someone wise, like for instance Seneca, he would advise you to find the wise teacher (he actually does this in the letters to the novice, Lucilius). But where does the wise teacher come from? Dear Seneca, you confront here either circularity or endless regress. Willliamson notes a mild form of the problem and offers an optimistic conjecture:

> We sometimes have a high enough level of expertise to bootstrap ourselves to a higher level of expertise by mutual criticism without input from anyone already at the higher level. Ms, 18 <u>*Metaphilosophy*</u> 42 (3): (2011) 215–229

But how do we arrive at the first level? In discussion at a conference Williamson has warned me that no good results can follow from completely misleading initial opinions (he avoids the term "intuition"). I agree, and am ready to believe that novices do have opinions-intuitions that are not hopeless. But this is only a start. How can initial opinions-intuitions get better? We should distinguish the principled from the applied, pedagogical matters. The principled issue is the following: how can the road from the ignorance of the novice to the achievement of expertise start and proceed? Once there is some initial level there, we can "bootstrap ourselves to a higher level of expertise by mutual criticism", without external input, as Williamson suggests.

The pedagogical level is the one we, philosophy teachers, encounter in our daily practice: how can we prompt the novice to make the required steps, without forcing him to take the route we think she has to take? Can one preserve one's intellectual autonomy and nevertheless become an expert?

---

[16] To quote a recent title of the book by Inmaculada de Melo-Martín and Kristen Intemann published in 2018 by Oxford University Press.

6.2.2 *Stages Defense—A Proposal*

Here is the sketch of my proposal, a defense that can be combined with others, but which might point to important routes that we have to expertise. I want to start from my mental model view proceeding in particular from the eight stages that my view discerns, and briefly propose this additional form of defense. Writing about X-phi Alexander (2016) has stressed the importance of "the mental model view", as he calls it (2016, Sect. 2), quoting as his main source Nersessian (2007). and also mentioning my work in references; he thinks the view is highly relevant for X-phi since if offers a plausible account of how philosophical intuitions are generated and of the role of imagination in the generation. (2016:60 ff.). I agree, and will try to show its relevance here.

TEs have each a wider biography, and reflection on these biographies might help. So, the leading advice is: consider the wider context, history into which the interesting intuitive answer fits, and their relation to stable intuitions. Look first at big or macro-TEs: they proceed step by step, like Plato's *Republic* and Rawls' Veil-of-ignorance. But smaller ones also have their biographies. We noted in chapter Two that often a person arrives at an intuition (e.g. linguistic, epistemic, or moral) thanks to the thought experiment, and others do the same. Sometimes, they compare their intuitions and discuss them between themselves. Professionals do this obsessively. They might agree or disagree and a pattern of history emerges.

Attention to stages will tell you how intuitions, or thinking based on them, can develop and progress. It could contribute to alleviating worries, and in fact situating the TE-activity within an acceptable framework of understanding of rationality. It could offer an idealized model for "becoming expert", avoiding circularity and regress. And it could serve to systematize already available local answers to local worries. So, we should look at stages.

Finally, I think one should use different medicine at different stages, since the ailments are different.

Consider first the early stages. At *stage one,* the stage of understanding the question, the question is often misunderstood. When this happens, it can be made clear, and especially the sub-issue of what is relevant can be made explicit. Also, the issues of motivation often trouble the unexperienced reader: why is Smith going through all this moves, replacing "Jones" with "person with ten cents in his pocket"? The point should be made clear.

Of course, this still demands an expert to do the teaching work. But we might imagine a more naïve person, reading the description of the Gettier case, and asking the question herself. But, if sufficiently motivated, the person can arrive at the proponent's motivations: "Oh, Gettier wants me to see the problem with "person with ten cents in his pocket", I see!" When I started doing analytic philosophy, in the midst of a very continental French environment, I had no expert to help me; I had to do all the guessing myself. We shall return to the understanding of proponent's motivation below.

At *stage two,* the experimenter's tentative building of the "model" can be helped and controlled.

We pass now to intermediate stages. Stages *three* and *four* (unconscious additional production and control by the relevant competence at unconscious level at stage three plus, at the next stage, the immediate, spontaneous verdict, often non-conscious, what we called "intuition core") are mostly sub-conscious and cannot be influenced directly, but some indirect control might be possible. *The fifth stage* covers explicit intuition.

How can deficiencies we encountered be mitigated at these stages? Better understanding of the situation presented by the proponent can prompt better modelling of it; that might help a bit against unreliability, by excluding some misunderstanding that make TE unusable. The subject might continue to be dogmatic and parochial, but some sources of irrelevance will be filtered out.

The stage six encompasses comparing, generalizing and intuitive induction. Here is an illustration for the later. Suppose an intelligent and thoughtful novice is considering the trolley and the fat man scenarios. She first reacts differently to the two: she would sacrifice a life in the first scenario, but she would not sacrifice a life in the second. And she might start wondering about where the relevant difference lies. Here, collating TEs and performing intuitive induction might help the novice, especially to counter the opacity of each intuition taken in isolation.

The richness of parallel cases might diminish unreliability a bit. Also, some dogmatism might be tempered. For instance, the novice's insistence that turning the switch is OK, but pushing the fat man is not might diminish once she sees the analogy between the two. She might move from "turn the switch" to "I don't know what I should or would do". Of course, one might be skeptical about this. For instance, my colleague and friend Klampfer claims that.

> /E/pistemicaly, the most controversial step, or, shall I say, leap is from a particular judgment ("such and such imagined/hypothetical social arrangement would (not) be fair/just", or 'such magic ring would turn me into an unjust person') to the conclusion that no social arrangement that shares all the relevant universal features can be just and, finally, that this is a matter of necessity, not simply a contingent feature of social world. Klampfer (2017):333

And he continues (referring to "political TEs" as to "PTEs"):

> Psychologically, we seem to find this transitions fairly easy and natural to make, but what, if anything, warrants them? What are the epistemically relevant features of PTE-generated intuitions? They may share their phenomenal properties with other TE-generated intuitions, but do they so clearly share their epistemic status as well? (ibid.)

But do we have to be that pessimistic? Either people are utterly incapable of moral generalizations, from real as well as from imaginary examples. But then, no moral rules can be constructed. Or, the critic thinks it's only generalizing in TEs that is problematic, but then the burden is on his shoulders. So, some optimism might be reasonable.

We then arrive to late stages that have to do with reflective equilibrium. *Stage seven* is the search for and finding of narrow reflective equilibrium. The isolated character of the intuitional judgment is mitigated by the work done in reflective equilibrium, where the given judgment gets a numerous company from other TEs.

At *stage eight*, the general knowledge of more empirical kind is brought into play. Traditionally, since Rawls, this stage is considered final, with good reason; it connects one's intuitional beliefs with everything else she knows or believe she does.

Here three mutually connected positive processes might take place. First, the equilibrated judgments might become more reliable. Second, dogmatism might further diminish, through confrontation with other judgments. And thirdly, if one is lucky, the inter-personal reflective equilibrium might bring the subject to discuss her views with colleagues from different groups; I might start talking to my two students from Hong Kong at Central European University. And I might discover a slight contrast in the way they judge the role of knower's authority. So, parochialism starts vanishing. This is a very optimistic, quasi ideal scenario.

We now proceed more cautiously, considering each stage in more detail. So, let us begin with the stage one. We shall argue that *it is way more important for our understanding of intuitions than philosophers are usually ready to admit.*

So, to the stage one, its dangers and possible remedies (with apologies for brevity). The main source of trouble is that subject's understanding of the problem influences her 'evaluation'; if she misunderstood the problem, either completely, or partly, by following a different reading of it than one assumed by the proponent, the final answer, and the 'evaluation' it contains will be deficient.

For warming up, let us start with the problem of cultural variations in meanings/concepts. Here one could, to some extent, accept the extant defense due to Ernst Sosa, who insists that such variation "need not be substantive" and might be merely verbal (2007:232).

> "maybe the term" knowledge is ambiguous like the term "bank."
>
> The fact that we value one commodity, called 'knowledge' or 'justification' among us, is no obstacle to our also valuing a different commodity, valued by some other community under that same label. And it is also compatible with our learning to value that second commodity once we are brought to understand it, even if we previously had no opinion on the matter. (Sosa 2009:109).[17]

As formulated, the example is quite incredible: if the word "knowledge" were ambiguous in radical sense, like "bank", it would translate into unrelated terms; (money-)bank is "Bank" in German, "banque" in French and "banka" in my native Croatian, whereas river(-bank) is Flussufer in German, "berge" in French and "obala" in Croatian. The linguist Anna Wierzbicka has recently published a paper on the topic in a collection dedicated to Gettier problem (Wierzbicka 2018). She points to interesting, but not dramatic differences, in stark contrast with the "bank" case.

Of course, if differences are less dramatic, this only speaks in favor of a moderate pluralism of meaning-concepts, and I think that we as philosophers can live with this.

Let us now pass to the dangers coming from the ordinary pragmatics of conversation, in particular from conversational implicatures. Remember Gettier's story of

---

[17] For more recent development see Sosa (2017).

Bob and Jill; Weinberg, Nichols, and Stich have in their (2001:29) retold it for the experimental subject:

> Bob has a friend, Jill, who has driven a Buick for many years. Bob therefore thinks that Jill drives an American car. He is not aware, however, that her Buick has recently been stolen, and he is also not aware that Jill has replaced it with a Pontiac, which is a different kind of American car. Does Bob really know that Jill drives an American car, or does he only believe it?

And they offer two answers concerning Bob: Really Knows versus Only Believes.

Now, as Cullen (2010) and David (2017) have noted, we are told that because of seeing Jill driving ": "Bob (...) thinks that Jill drives an American car." But in ordinary conversation, claiming that a person "thinks that..." implicates that the person does not know it. So, the subject is being primed to respond "only believes", not on the basis of her impartial judgment, and of reasons Gettier thought of, but on the basis of conversational suggestion stemming from the use of "thinks". (See also Kauppinen (2007)). The example is not particularly dramatic, but it points to the force of pragmatic factors in the understanding of the task.[18]

What about mitigating the problem? Here, the task is with the experimenter, the proponent. If she really wants to test the novice's spontaneous reaction, she should avoid wrong suggestions-implicatures. In the Bob example, the claim that Bob thinks suggests Bob not knowing. If the proponent said "Bob is therefore convinced that Jill drives an American car ", there would be no such implicatures; being convinced is conversationally neutral between knowing and merely believing.[19]

Staying with the Gettier scenarios, we next pass to motivation of participants, both the proponent and the subject. Here, as most philosophy teachers have noticed, novices have difficulties with getting the point of the scenario(s) proposed.

The typical scenario is the Smith/Jones story. The guiding question is probably: why should I worry whether Smith *knows*? He is right about the person with 10 cents in his pocket, the person, he himself, will get the job, he has good reasons for his

---

[18] For another kind of pragmatic considerations see Sytsma and Livengood (2011). See also Adams and Steadman (2004) and Driver (2008).

[19] The issue of understanding is now being made prominent in several lines of discussions, that go beyond the discussion in experimental philosophy, where theorists often use procedures from the methodology of psychology in order to control variables that might influence subject's understanding of the question.

For logical intuitions the famous issue is subjects' understanding of everyday words for logical constants, most famously "if...then"; is it conditional and biconditional and if the first, what kind of conditional (see chapters Four and Five of Stenning and Lambalgen (2008).

Concerning arithmetical intuitions there has been a recent intriguing debate about a Brazilian Indian tribe whose members were described as being unable to place objects from even two small bunches in one–one correlation: the researcher took it as a very significant result since the tribe also seems to lack numerals/number words. Others accused him of simply having misunderstood his subjects. See Everett and Madora (2012) For linguistic intuitions a famous, not to say notorious issue is whether illiterate subjects indeed have grammatical intuitions at all see Schütze (1986).

conviction, so what does this additional insistence on knowing as opposed to merely truly and justifiably believing mean?[20]

But there are more questions about the motivation behind the proposed story. What's the point? What is this business with S counting coins in Jones's pocket? Why does Smith reason from the important info that Jones will get the job to the rather irrelevant thought that person with 10c in his pocket will get it? Why would anyone do it? Why is it important whether S knows the irrelevant fact that 10c person will get the job? The expert has gone through this kind of traps thousand times, and doesn't waste a second on them; the novice is, completely normally, dumbfounded by them.

The combined issue of understanding the proponent (why am I being asked about this?) and understanding the motivation of the principal character (why would anyone do this?) might culminate in the first-person motivation question: Why would I go along and simulate Smith's strange behavior (coins) and reasoning (from relevant to irrelevant info)? In addition the trouble with understanding in the given situation may increase with any kind of cognitive overload that may accompany it: in the seminar, the tiredness of the student, and so on.

How can we mitigate the problem? First, the proponent might come up with motivationally easier example. Here is the story I used with novice students, mostly female ones:

> Suppose your partner is cheating you. You have evidence it's with Jane. You believe "I am being cheated". In fact, he is cheating you with Kate. Do you know you are being cheated?

Students understand immediately, and are subsequently able to project it onto Smith-Jones story and grasp the original Gettier example. Why? First, they feel immediately that it is important for the person in question to *know*; no lesser epistemic status will do. Second, there is no puzzle of irrelevant seeming steps, like introducing the "person with 10 cents" description. Third, the scenario is easy to grasp, and the possible consequences easy to imagine (I start hating poor Jane, and she is completely innocent).

Once the students see the basic structure, I tell them that philosophers have a similar, but famous query; there, being cheated is replaced with getting a job, and the person to get it is characterized in a somewhat funny way, with the possession of ten cents. But the changes do not influence understanding any more. Why? Because in the jealousy version the problem is of immediate emotional relevance to students, quite motivating to unpack, and easy to imagine/ simulate. The perfect ease of understanding and simulation diminishes the cognitive burden. Further, the subject is presented as reasoning to the most important fact (I am being cheated). This makes it easier to realize that the coincidence (Jane-Kate) is accidental and misleading. This then provides the understanding of the structure of the Gettier-problem.

Sorensen is right to stress the importance of acquaintance with a range of similar cases: "The expert also has a holistic advantage over the novice. The expert is familiar

[20] See Baz (2012). Baz in his paper also has interesting things to say about the ways semantic/pragmatic distinction is relevant in the Gettier context.

with a constellation of comparison cases. Whereas freshmen evaluate a Gettier case in isolation, the epistemologist considers the case against a background of similar cases (…) Comparative, contextualized judgment is more reliable than absolute judgment." Sorensen (2014:141). Here, the subjects get a similar, but more motivating case, and it makes life easier for them. The contrast is clear. The novice-layperson facing the Smith-Jones scenario encounters multiple external problems: first, understanding the issue, understanding (simulating) Smith's motivation and actions, physical and epistemic, and finally the issue of getting motivated to judge them taking all things into account. These complications once removed, the competence shines through! For the expert, on the contrary, the external sources of problem have been for the long time blocked, so she is easily concentrating on the crucial point (no cognitive overload), and is highly motivated to answer the question taken to be crucial to the whole profession.

Suppose the novice has become minimally trained, either by an expert, or, on our optimistic scenario, by one's own effort. She would become sensitive to pragmatic effects, she would better understands the problem (the content) but also better understands motivation (specifically the philosophical one). Finally, a more difficult question: could a novice do it herself? Suppose she reads the Gettier story all by herself, no teachers around, and she is independently motivated to understand it. She has all the time she needs, unlike the subjects in X-phi tests. "What is this business with 10 cents in the pocket", she asks herself. "Well, the author needs a belief that will be true both of Smith and Jones, but Smith will believe it for the wrong reason. I see, he is presenting me with something that is believed by Smith that is true of both guys, and is believed by Smith for a good reason, but completely unconnected with its truth!" Could an intelligent novice do it? If yes, the problem is mitigated. I leave the judgment to the reader; I am myself optimistic. This is as far as one can come with the stage one.

We now pass to the intermediate stages. First, we have the stage two, tentative conscious production. I suppose it involves modelling, and we shall look at the typical case where the subject is building the model of the scenario proposed. This stage, and those that follow, all the way to the stage six, involve subject dealing with a particular case, in isolation. "Imagine you are on a super-speedy train, and there is a signal coming from the front side of the vagon…", or "Imagine Smith worried about the job and reasoning about Jones" Imagine an agent in a completely deterministic universe, murdering his wife and kids!" No variation, until the stage six. The particular problems here come from various sources, and we shall abide by a typical one; others are to a large extent analogous. Our typical, very broad kind involves affective matters. We have mentioned the bad influence of affective surroundings in the first section, and also the problematic role of affective content. Remember the example of a murder performed in a deterministic universe, where subjects typically assign responsibility, neglecting the deterministic context (subject don't assign responsibility for neutral deed).

Interestingly and fortunately, these affective influences have been discussed in cognitive psychology in general, in context wider than TEs. The common term for the topic is "bounded rationality", and the most interesting fact about the studies is that leading authors proposing the thesis that our rationality is bounded, are way less pessimistic about the consequences than the author from negative X-phi paradigm. Let me mention two leading authors, each followed by a suite of collaborators, Kahneman (2011) and Dan Sperber (see Mercier and Sperber (2017). Kahneman is generally more optimistic about ordinary human cognitive capacities than the negative X-philosophers and Sperber is also much more generous to moral intuitions than the X-Phi group.[21]

Let me start with more holistic dual processing proposal of the Kahneman style. It proposes the division between quick and slow thinking processes. Then, it suggests that in our cases the quick and dirty processes simply substitute the question: for instance, the question of blame & punishment replaces the question of responsibility pure and simple. In such affectively colored TEs the intervention of the wrong mechanism is to be expected. And the more modular theories, like Sperber's, would claim that in our cases the wrong module grabs the issue and promotes the irrelevant features of the situation to central role. Here, as Haidt (2001) puts it, the emotional tail is wagging the rational dog.

Priming, say by order of presentations (which we illustrated with Truetemp variations) induces the same phenomena. Finally, the proposal in Le Buon et al. talks about two sub-components in the example at hand some kind of moral competence competes with affective sub-component. Unfortunately, the later wins. The result can then be described as performance error. This offers support to the general idea that important mistakes derive from the wrong instance taking over in thinking about the issue at hand (idea to be found in Kahneman, Sperber, **etc.**).

Here is a quick, and to my mind impressive solution proposed by Ernest Sosa.

> /…/ we may then find different "interpretations" at work in the verbal disagreement between the affect-affected intuiters (who react to the specific description of the dastardly deed) and the cold theoretical intuiters (who respond to the abstract question of whether any agent can be responsible in D). Of course, it remains to be seen why the one concept is more readily engaged by the affect elicited with the specific case, and the other more readily by cold reasoning about the abstract issue. (Sosa (2007) pp. 99–107, p. 104)

When confronted with the question whether the agent in the story is responsible for his deeds the affect-affected intuiters deploy an affectively colored concept of responsibility, tied to blameworthiness, dastardly character and the like. The cold theoretical intuiters stick to the abstract, philosophical concept of being responsible. The sentence "The agent in the story is responsible for his deeds" then expresses two contents, and the affect-affected intuiters react to the one and the cold theoretical ones to the other. There is a difference of the topic, and no disagreement about the same topic.

I like the proposal very much, but I am still worried about the X-phi question: if folk intuitions are merely about affectively colored concepts, there is no continuity

---

[21] An earlier attempt to compare X-phi and theories of bounded rationality is to be found in Buon et al. (2013) "A non-mentalistic cause-based heuristic in human social evaluations", *Cognition* 126 149–155.

I have profited a lot from discussion with one of the authors, Pierre, and I thank him for this!

about ordinary and philosophical intuitions, how can the philosophers appeal to the former in order to defend the later. Obviously, they cannot. But then, negative X-philosophers win. So, we shall have to look for further solutions.

Back to the process in which the problem is generated by the takeover from the wrong instance. At what stage does it take place? Sometimes, at the stage two: the wrong instance builds, or helps build the model,

In other experiments, especially those where the scenario is detailed and very clear, probably not at building of the model: the model is dictated by the scenario itself. In these cases, the probable stages of intervention of the wrong sub-system or module are the next, unconscious ones. We shall concentrate upon the obvious candidate: the stage three, the main locus of the unconscious processing of the problem.

Two important things may happen here. First, the takeover itself: an instance (a module, wider sub-system) takes the problem in its hands. Second, this results in the unconscious production (inference) and control by that instance (at unconscious level. The takeover is the ideal place for the intervention of the wrong instance, and cognitive scientists mention possible causal explanations; the paper by Buon et. al. offers us the overload as a possible cause.

They also propose that it is the performance error. The modules or sub-systems are reliable, their competence is not problematic, the irrationality in the case at hand comes from circumstances of performance, like the overload. Now, if the overload is the main cause, this points to two interesting things: first, that a part of the reason that experts are better than novices is simply that they find the task easier: the expert has seen a lot of analogous cases, doesn't have to reflect consciously about every detail, and has a clear picture of the goal of the proposed TE. So, no overload for the expert, so, the right sub-system (= competence) does the job, so, no performance error.

If it is then, no need to say, it is a good news. Why? First, because it makes philosophical competences parallel with linguistic and mathematical ones. There, there is a whole lot of performance error with beginners; and if linguists and mathematicians (and math teachers) can function with it, we philosophers can as well. There is no need to go to skeptical extremes, Stich-and-Kornblith fashion. Instead, we need simple ameliorative steps, like making the task easier! For philosophy class, it means breaking it into segments, going over it several times, and so on (if you teach philosophy, you already know it). We may ask linguists and the math teachers for advice, they have been successful with it.

- So much for the general remarks.

We quickly pass to stage four, the immediate, spontaneous verdict, often nonconscious. One obvious source of danger derives from the isolated character of the judgment: the competence just outputs the judgment with no info for other modules about its origin, and reasons for it. It is "a 1-bit info", as Weinberg puts it in a related context (2007:330).

But what about the sub-stage that we hypothesized primarily for the modular theory after the stage four, pure and simple, the additional unconscious processing by Central Processing Unit? (The analogous process for the slow-and-fast, two systems theory would be the slow system taking over, and checking the output of the quick one. In this holistic option the result of stages two and three is fed directly to the conscious system.) Cannot the Unit or the conscious system help? Since at one of the previous stages the irrelevant elements have won over relevant ones, the result is bad. But, the central Unit, has no indications of the possibility of mistake, so, in default case, does not react. Even worse, additional irrelevant motives might win ("I want to inspire respect, so let me show that I am a resolute, manly person and I can do this by immediately condemning the agent in the determinism story; I don't care for details about determinism, I care about my image in the actual conversation.)

This brings us to stage five: immediate spontaneous answer, i.e. the conscious intuition. Mitigating it can be done only on previous stages, so let us remind ourselves of what can be done.

Better understanding of the situation presented by the proponent can prompt better modelling of it; that might help a bit against unreliability, by excluding some misunderstanding that make the TE unusable.

The subject might continue to be dogmatic and parochial, but some sources of irrelevance will be filtered out. Can we counter the Argument from unreliability? Not quite, but some mitigation is possible. We agree that.

1. "Unreliable judgments are severely deficient from an epistemic point of view."

The challenge is formulated in the next premise:

2. Judgments elicited by most of the philosophical cases that have been examined by experimental philosophers are unreliable.

Well, an effort at early stages can make result more reliable; not fully so, but better than initially. One contribution might come simply from better formulation by the experimenter, formulation that is able to block pragmatic misunderstandings, and to make clear the motivation of the TE for the subjects. Others can come from more practice of the subject with TEs; not 10 000 h of reflection, but a few hundreds.

Machery has rightly noted that "we ought to refrain from making a judgment (…) when most judgments of this kind are plausibly severely deficient from an epistemic point of view (2017:103, premise 4 of the Argument from reliability). My optimistic suggestion is that we might start moving from "severely deficient" judgments to judgments that are still not perfect, but are less than severely defficient. No need to suspend judgment.

But the life of intuition is not over.

What about Dogmatism? Hopefully, the processes we discussed can mitigate it. The Parochialism of the subject stays, but it is not supported by completely irrelevant considerations, as it might have been initially.

Stage six, varying and generalizing is a turning point, quite important and potentially helpful. Chisholm has used the term "intuitive induction", borrowed from Ross's translation of Aristotle, to give it a name. Many philosophers, from Aristotle to Chisholm and J.R. Brown have stressed its importance. It has not been much addressed by X-philosophers. Intuitive induction is supposed to culminate in stage seven, the general belief at reflective (personal) level, a kind of general intuition.

We shall consider both stages together.

Intuitive induction might offer possibilities of amelioration and correction. Note that the most dramatic results negative are obtained in separation conditions (like in other (ir-)rationality studies). Showing options together usually helps; the rule is valid also outside phil.: Kahneman gives examples from business decision making.

> The features that caused the difference between the judgments of the options in single evaluation—the poignancy of the victim being in the wrong grocery store and the anchoring on the prize—are suppressed or irrelevant when the options are evaluated jointly. (Kahneman 2011:354)

The same for presenting options in a different order. Remember Truetemp-like cases of putative knowledge were apparently influenced by the presentation of contrasting cases, in particular by the *order* of presentation of the various examples (Swain et al. 2008). And, separation was crucial: one group gets one order of presentation, another group another.

The subject needs to compare, and changing order might change his/her resulting intuitions. Compare with experts, as described by Sorensen:

– The expert also has a holistic advantage over the novice. The expert is familiar with a constellation of comparison cases. Whereas freshmen evaluate a Gettier case in isolation, the epistemologist considers the case against a background of similar cases … Comparative, contextualized judgment is more reliable than absolute judgment. p. 141.

But the novices can come across comparison cases, and good authors do present TEs in combinations (e.g. Hume on parricide (thanks Isabel Patkowski)).

Here is my favorite example, the famous story from the Old Testament, extolled by authors like Lessing, and brought to bear on philosophical issues by Tamar Gendler: Nathan 1st versus 3rd person.

There were two men in one city; the one rich, and the other poor. The rich man had exceeding many flocks and herds: But the poor man had nothing, save one little ewelamb, which . . . grew up together with him, and . . . was unto him as a daughter. And there came a traveler unto the rich man, and he spared to take of his own flock . . . but took the poor man's lamb, and dressed it for the man that was come to him. When David hears this story, his "anger [is] greatly kindled against the man."

He holds the man to be deserving of disapprobation and punishment, and says to Nathan:

> As the Lord liveth, the man that hath done this thing shall surely die. And he shall restore the lamb fourfold, because he did this thing, and because he had no pity.

At this point, the circumstances have been set for the delivery of the punchline. Nathan says famously to David:

> Thou art the man. . . thou hast killed Uriah the Hittite with the sword, and hast taken his wife to be thy wife, and hast slain him with the sword of the children of Ammon. . .

With a shock of recognition, David reframes his understanding of the circumstances in which he has placed himself, and says to Nathan: "I have sinned against the Lord." Szabo (2010:130). With the variation and the resulting general belief, that now has a wider and recognizable basis (no "dark margins") we have principled results, for example, enhanced resistance to order of presentation and similar kinds of priming. The pedagogical application is immediately obvious.

Suppose now that our subject has, like the king David, become wiser due to efforts of comparing cases. Return to initial stages creating trouble. For instance, stage two. The unconscious processing has been re-trained; no substituting the question (in determinism example, look at responsibility pure and simple and suppress irrelevant features). Priming, say by the order of presentations (Truetemp, etc.) has been blocked. It is easy to translate the result to pedagogical context.

The same with stage three. Subject's cognitive system is now operating on relevant propositions and reasoning from relevant features. Unconscious (empirical-?) theorizing by central processing unit goes well: relevant motives win (the "I want to inspire respect," type of motivation is weeded out). In the holistic options relevant elements win over irrelevant ones.

Problem: it is all merely pedagogical. Could David have done without Nathan? In the principled version, he would have to go much deeper in self-questioning. For instance:

> Well, I have now Bathsheba, I have taken her from her lawful possessor Uriah and I have sent him to death. Now, the rich man has taken away the beloved sheep from the poor man, and served it to his guests. My, there might be some similarity here! Well, the rich man has not slain the poor one, just his sheep; I sent to death Uriah, and not Bathsheba. But, still, I am dead certain that the rich man did an awful thing in depriving the poor one from the creature he loved. And I deprived Uriah from the creature *he* loved.

It is not psychologically impossible that someone, even a military hero, might go through this kind of reasoning. And arrive at revelation: "I am no better that the rich asshole!".

We already said that the richness of parallel cases might diminish unreliability a bit. Passing from atomistic to molecularist patterns helps a lot. So, suppose that after some self-training, or after a bit or clarification by the proponent, at early stages the unreliability of the subject has become less dramatic. Now, comparisons and generalization in self-training, with no Nathan around (or training, if he is available to the king) pushes it further down.

Also, dogmatism might be dramatically tempered, if one is lucky. We noted that showing options together, with variation in order, and so on, does enhance the rationality of the decision-making. Parochialism remains, but now it has been cleared from a lot of irrelevance…

Interestingly, variation can be used by proponents of TE for the opposite purpose, to show that no changes, no adaptations will save the theory tested from catastrophe. A famous example is Heisenberg's microscope TE from 1927. (See for example Peacock (2018:225 ff). Suppose you want accurately to determine an electron's position and momentum using microscope; you hit it with a photon particle, and locate the position of the particle after it has been reflected by the electron. Unfortunately, for greater accuracy, one only needs a microscope using light of smaller wavelength, and therefore higher energy. But the energy will be to big; the variation in the acquired momentum of the electron will become non-measurable. Then, its position will be clear, but its momentum will escape measurement. Or, vice versa, if you hit it with low energy particle, the low energy will be associated to bigger wavelength, and then position gests lost. Translated into our terminology for TEs, there is no possibility of a successful intuitive induction from particular outcomes. Back to our main topic.

As we noted above, *stage seven* is the search for and finding of narrow reflective equilibrium. The isolated character of the intuitional judgment is mitigated by the work done in reflective equilibrium, at *stage eight*, in particular the sub-stage of looking for and finding wide reflective equilibrium, where the general knowledge of more empirical kind is brought into play. Traditionally, since Rawls, this stage is considered final, with good reason; it connects our intuitional beliefs with everything else we know or believe we do.

So, how can reflective equilibrium help? First narrow, then wide (intra- and interpersonal)? In the narrow intra-personal reflective equilibrium the subject combines several micro-TEs, or many micro-TEs into a larger whole. So, suppose our subject Jane has put her thoughts in order, she has a coherent picture of the issue. In what ways does this help? Putting one's intuitions in order accomplishes the work started with comparing the cases (stages six and seven). The case under consideration is not only compared with its relatives and its contrast class, but connected to topics that go with it: possible causes, ranges of consequences, and so on.

Hopefully, this yields explicit understanding; now Jane understand the whole. Of course, there is always the danger of confirmation bias; what if Jane looks only for causes that fit her evaluation of the target proposition, suppresses the rest, and the like. If she is focused on the evil of killing one's partner, she might suppress considerations of other kinds of deeds necessitated by determinism, abstract from the impossibility for the agent to act otherwise, and so on. One might say that her implicit understanding of the scenario forces her explicit understanding in narrow reflective equilibrium.

Or, in culturally determined intuitions, the cultural whole would make its appearance at the level of reflective equilibrium. What is to be done? The natural move is to pass from reflective equilibrium limited to one person to an intersubjective, collective reflective equilibrium. Some authors, for example Sperber, would recommend that Jane discusses her intuitions with others. Thus, her epistemic vigilance is increased, and she starts reflecting about every sentence she hears. He claims that argumentation dramatically improves our thinking and is adamant about this, (there is a lot to be said here, but i must be short).

So, suppose Jane discusses the issue with her friends. Initially, they disagree.

Then, one of the two things happen. Their opinions polarize, and they never reach a reflective equilibrium at the group level. (bad luck—Sperber has been insisting upon dangers of polarization). Or, their opinions converge and a common picture is built. The authors we have looked at have suggested several ways to make things better once the disagreement have been noted. For instance, assign a role to each kind of intuitions that makes them acceptable. To stay with ascription of intentions and responsibility assume that in the case of the company boss, or someone pretending to be a boss, ascription of intentionality is gudied by need to praise and blame and responsive to different conditions for praise and blame.)[22]:

Second, accept that some intuitions are incorrect, but argue they can be explained away, for instance by the appeal to competence/performance distinction) and do not throw negative light on other intuitions.

Third, argue that one can in the long run improve one's intuitions (as done in rationality studies, by authors like Kahneman (2011)).

For cultural differences one might need intercultural dialogue, indeed of collective inter-personal narrow sort. The contrasting groups, or their representatives, now discuss matters together and decide a non-parochial solution. (I did it in a seminar with students from Hong Kong, it was interesting and rewarding).

Philosophers are aware of the famous historical versions of such philosophical search for reflective equilibrium. With Socrates it is the discovery of the dialogue that is a crucial example. On the contemporary scene one has a contrast: on the continental side less attention to argumentative discussion, more to the poetic rhetoric of persuasion, on the analytic side a lot of discussion, but also the danger of endless hair-splitting. Another evidence in philosophy in the last century and a half is the growth of commentaries and analyses, common to analytic and continental traditions (perhaps anticipated in medieval scholastics). More comprehensive understanding is the goal, and collective reflective equilibrium the desired state.

Of course, one should not exaggerate. Tristram McPherson present such an exaggerating philosopher, and then argues that the character will never reach his or her goals.

> In light of its extraordinarily idealized character, wide reflective equilibrium is not a state that an ordinary human could ever be in. This precludes at least direct investigation of this state by scientific psychological methods. This in turn might seem to mitigate the puzzle mentioned above: perhaps social scientific investigation of the target facts is not a serious competitor to Rawls' method, because it is not possible. McPherson 2015:660[23]

But there is no need to go so far with the demands.

Back to the intra-personal level for a moment. The next step in ameliorating my intuitions is bringing my wider knowledge to bear on the issue. What is science telling me about determinism? If I listen to scientists I might discover an unpleasant piece of news: we do live in a completely deterministic universe. So, what about

---

[22] This is, of course, just a beginning. There has been much work done on compatibilism and incompatibilism as represented by folk intuition; we cannot go into it here. See, for instance Nahmias et al. (2005, 2006), and Chan et al. (2016).

[23] In short, in light of suggesting that the ideal moral enquirer consider all reasonable arguments, the method of reflective equilibrium fails to Adjudicate the central controversies within moral methodology. Rather, it in effect absorbs all such controversies, as inputs. This undercuts any claim that the method provides an informative answer to pressing methodological questions in ethics. Note that this is true even holding as fixed Rawls' project of moral theory; how we should proceed if and when we attempt to move from moral theory to actually defending answers to ethical questions is even less clear 2015:661.

responsibility? Suddenly, the responsibility for trivial positive, trivial negative and non-trivial terrible and dastardly deeds has to be judged together.

Again, I might choose to discuss it, say with my physics and psychology friends. And with them, I attempt to reach inter-personal wide reflective equilibrium. The crucial role of such wide reflective equilibrium consists in the fact that different kinds of expertize on various domains are brought to bear on the initial judgment. Even theories of intuition, naturalistic or not, are brought to bear.

It is time for a summary (with apologies for sketchiness and brevity). Ours is a modest proposal: some intuitions might be saved. It is not a mere expertize defense, and it insists upon distinguishing the principled from the applied, pedagogical matters.

It notes that dangers and promises are unequally distributed in the biography of a typical philosophical intuition: dangers abound in early stages, promises come in later. Collating TEs and performing intuitive induction might help, especially with the opacity of each intuition characterizing it when it is taken in isolation. Then individual reflective equilibrium often caters for coherence, in good case, and collective narrow reflective equilibrium might mitigate the differences. (might, if one is lucky; if not, one gets polarization and negative spiral). Thus, one recognizes then the margins of one's own reliability intuiting. It all gets strengthened and integrated in our general knowledge in the collective wide reflective equilibrium. So, on the principled level, we have hopefully, avoided the problem of regress or circularity, and shown how expertize is possible.

We have also brought together the normative epistemological. and cognitive accounts, having sketched the connection between issues of TEs and of rationality. Cognitive theories of bounded rationality nicely reproduce and generalize the experimental results, but also suggest a more optimistic stance towards the results. Some negativistic X-philosophers might not be happy: they argue that intuitions, in particular the philosophical ones are especially bad.

But we have shown that in the course of reflection, the subject, having gained a certain amount of sophistication can re-consider the initial stages of her thought experimenting, and mitigate the harm in the ways we already sketched. Theoretical illumination might be the final result!

6.3 Conclusion

In many areas TEs are extremely useful heuristic devices, as illustrated by famous physical TEs. In philosophy, their role is even more prominent, both in making explicit the deep structure of our understanding of the world, and in guiding our normative judgment, in a number of areas, like morals, aesthetics, political thinking and epistemology. This makes them indispensable for philosophy, and we have to find ways to make them better and more useful. The main challenge to TEs has come from experimental philosophy, and in this chapter we have tried to summarize it and to indicate ways in which it can be answered.

Experimental philosophers have been rightly stressing the importance of including scientific psychological research into the discussion of TEs. The negative school has drawn pessimistic conclusion from their efforts. Here, we have defended the TEs, and appealed to widening the cooperation with scientific cognitive research in the direction of including the dominant theories of bounded rationality which show that elements of irrationality found in TEs are common across all domains of thinking and reasoning; they also suggest more optimism than is present in the negative X-phi program.

The distinction between principled possibility and the actual institutional (pedagogical) work is crucial for plausibility of the defense. We have here tentatively proposed one particular framework for addressing the issue. Instead of simply ascribing bounds to the overall rationality of experimental subjects, we proposed to consider different problems as they arise at different stages of a typical TE. Once the type of the problem is located at its particular generating stage, it is easier to find a more local medicine.

Here is the hope tied to stages defense. In the first stages the way in which the subject understands the question often determines the rest, and ill-understood or misunderstood questions might at the end of the day result in theoretical disasters. Careful formulation of the task that lead to understanding of what is required somewhat *diminish the unreliability*. And this in spite of the atomic, isolated character of intuitions considered. Dogmatism and parochialism for the most part stay, but cleaned from completely irrelevant components. Note that on the principled level the process works with minimal external prodding, and all the work is done by the subject. The morals of these debates for philosophers is very clear: *the first stage is way more important for our understanding of intuitions than philosophers are usually ready to admit.*

Many dogmatic pronouncements on the nature of intuitions may reside upon simple ignoring of the complexities of this first stage. Since experimental philosophy is paying a lot of attention to cognitive research on intuitions (and sees itself as forming a part of such a research), it is reasonable guess that the first stage will get more attention of philosophers in the near future; both friends and foes of experimental philosophy will have to concentrate upon it.

In later stages, with comparison and generalizing the atomistic approach is abandoned. In consequence they *dramatically diminish unreliability*. What remains of dogmatism and parochialism is now *a bit more coherent*. With the comparison, variation and the resulting general belief, that now has a wider and recognizable basis we have principled results, for example, enhanced resistance to order of presentation and similar kinds of priming. The pedagogical application is immediately obvious. The process is in principle possible *in a completely autonomous setting*, with the subject doing all the comparing by herself. (In practice, we do learn from experts, but *the principled question is solved*.)

Suppose now that our subject has, like the king David, become wiser due to efforts of comparing cases. Return to initial stages creating trouble. For instance, stage two. The unconscious processing has been re-trained; no substituting the question (in determinism example, look at responsibility pure and simple and suppress irrelevant features). Priming, say by the order of presentations (Truetemp, etc.) has been blocked. It is easy to translate the result to pedagogical context.

Reflective equilibrium *blocks unreliability*. It can *dramatically diminish Dogmatism*. In the case of protracted and successful *inter-personal* equilibrium *the problem of Parochialism is solved*; the contrasting groups, or their representatives, now discuss matters together and decide a non-parochial solution. So, stage defense might help after all.

6.4 Are Scientific TEs Better than the Philosophical Ones?

Until now we have been discussing criticisms of philosophical TEs. What about science? Let us briefly turn to a debate in which the critics stayed with skepticism about philosophical TEs but defended the superiority of TEs in science. The debate took place at the beginning of the new millennium but has kept its actuality.

Peijnenburg and Atkinson note that the conclusions of TEs are often contradictory, and in some cases conclusions "beg the question: they hinge on intuitions of which the truth or falsity was supposed to be demonstrated by those very thought experiments." (2003: 310). Further, Peijnenburg and Atkinson have compared most famous TEs in the domain of philosophy and of science, e.g. Jackson's Mary and Einstein, Podolsky and Rosen's (EPR) one on quantum physics and entanglement.

Interestingly, the final verdict turns not on experimenting itself but on its available frameworks.

> Does Mary, in actually seeing colours, learn anything she did not know before? Of course she does!, answers Jackson (1982). Of course she does not!, counters Paul Churchland (Churchland 1985). Who is right? Again, there is no way to tell. This is not to say that Jackson and Churchland have not given arguments for their beliefs. They have, but only to shunt the disagreement to another level (2003:309).

And they bring a verdict:

> This debate is doomed continually to go round on a merry-go-round, since the participants more or less tacitly endorse two totally different starting points (2003:310)

On the side of physics, their favorite is EPR TE. Since I have not presented it in the book as yet, and it merits to be introduced, let me use their presentation, that will also illustrate their strategy:

> Suppose that two particles are created together in such a way that their positions and their velocities at later times remain correlated. That is, if one measures the position of particle 1 at one of those later times, t, one can calculate, and thereby predict with certainty, the position of particle 2 at t. A similar story can be told for the velocities: if at t one measures the velocity of particle 1, one can immediately infer the velocity of particle 2. Although one would have to decide whether to measure the position or the velocity of particle 1 at t – one could not do both—EPR stress that this decision does not affect particle 2. After all, the two particles could be miles or even light years apart when the measurement on particle 1 is

made, so that, although the measurement will disturb particle 1 and change its properties, no such disturbance could be instantaneously transmitted to particle 2 and change its properties. Since either property could have been measured for particle 1 and thus be predicted with certainty for particle 2, and since the ontological status of particle 2 does not depend on whether one decides to measure the position or the velocity of particle 1, it follows that the position and the velocity of particle 2 must exist:

'If, without in any way disturbing the system, we can predict with certainty (i.e., with probability equal to unity) the value of a physical quantity, then there exists an element of physical reality corresponding to this physical quantity'. (Einstein et al. 1935, 777) (2003:317)

And they note that "Einstein ties physical reality to absolutely correct predictions that could be made; Bohr ties it to measurements that actually are made" (Ibid.)

Now, the situation looks similar for the two experiments: endless disagreement of leading thinkers, and then a series of interventions into experiment, with proposals of change, into its immediate conclusion and the way to place it in reflective equilibrium with dominant views.

Our authors see matters differently. They note that, "although this thought experiment failed (as a thought experiment), the consequences were again not as disastrous as they were in the case of the philosophical thought experiments. In 1952 David Bohm retooled the EPR experiment in such a way that John Bell, in 1964, could make a prediction." (Ibid.) EPR was made testable, they say, Alan Aspect tested it and obtained negative result. In spite of parallels between philosophy and science, "in science the damage is restricted: in some cases we can still fall back on theories, whereas in cases like the EPR-TE we can arrive at finally making a real and decisive experiment.

Indeed, in another paper Atkinson has suggested that thought experiments in physics are of value only when they are related to or inspire real scientific experiments (Atkinson 2003:219).[24]

In his criticism, Cohnitz rightly notes that Bell's and Aspect's work "did not close the topic." (2004:12) The Aspect experiments are under discussion, he reminds us. On the side of philosophical TEs he reminds us that "the intuition triggered is that Mary learned something new! The intuition is not that physicalism is false. This only follows after adding other assumptions."(14) And he adds in conclusion that "whether a thought experiment is a good one then turns out to be a question we can never answer in advance. (2004:29). When they designed the EPR thought experiment, Einstein, Podolsky, and Rosen couldn't foresee Bohm's reformulation nor Bell's inequalities. EPR turned out to be "good" 50 years after it was designed. That is pretty unhelpful for methodological considerations" (2004:17).

So, what should we think about the two standpoints? First, note that the alleged minuses of TEs are not really minuses of thought experimenting as such, but rather deficiencies of available wider frameworks! Further, if an important thesis is scientifically testable in some reasonable time, then TEs teaching us about it can still be very useful. Note that issues that worried philosophers stopped being philosophical issues once they became solvable by science: the mystery of life and its varieties

---

[24] See also Rédei (2003).

was a typical philosophical question before the advent of the theory of evolution. I see nothing bad for philosophy in these developments. But what if it is not? What if there is a region of philosophical problems that science will not solve? This is an open possibility, and it will keep a space reserved for TEs only. But I am neither specially shocked by the possibility, nor attracted to it.

So, I propose a kind of equality of scientific and philosophical TEs; neither of the two kinds is especially problematic, and both are useful and promising.

References

Adams F, Steadman A (2004) Intentional action in ordinary language: core concept, pragmatic understanding? Analysis 64(282):173–181

Alexander J (2016) Thought experiments, mental modeling, and experimental philosophy. In: Nado J (ed) Advances in experimental philosophy and philosophical methodology. Bloomsbury Academic, London, pp 53–67

Alexander J, Mallon R, Weinberg JM (2010) Accentuate the Negative. In: Knobe J, Nichols S (Eds) Experimental philosophy: Volume 2. Oxford University Press, New York (2014), first published in Review of Philosophy and Psychology 1(2):297–314 (2010)

Atkinson D (2003) Experiments and thought experiments in natural science. In: Galavotti MC (ed) Observation and experiment in the natural and social sciences. Springer Verlag, Switzerland, pp 209–225

Baz A (2012) Must philosophers rely on intuitions? J Philos 109(4):316–337

Björnsson G, Pereboom D (2006) Traditional and experimental approaches to free will and moral responsibility. In: Sytsma J, Buckwalter W (eds) A companion to experimental philosophy. Blackwell, Malden, MA, pp 142–171

Buckwalter W (2016) Experimental philosophy, Oxford Bibliographies Online. Oxford University Press, Oxford

Buon M, Jacob P, Loissel E, Dupoux E (2013) A non-mentalistic cause-based heuristic in human social evaluations. Cognition 126:149–155

Chan H-Y, Deutsch M, Nichols S (2016) Free will and experimental philosophy. In: Sytsma J, Buckwalter W (eds) A Companion to Experimental Philosophy, Blackwell, pp 158–172

Churchland PM (1985) Reduction, qualia and the direct introspection of brain states. J Philos 82(January):8–28

Cova F (2016) The folk concept of intentional action: empirical approaches. In: Sytsma J, Buckwalter W (eds) A companion to experimental philosophy, Blackwell, pp 121–140

Cullen S (2010) Survey-Driven Romanticism. Rev Philos Psychol 1(2):275–296

Driver J (2008) Attributions of causation and moral responsibility. In: Sinnott-Armstrong W (ed) Moral psychology. vol 2. The cognitive science of morality: intuition and diversity. MIT Press, Cambridge, MA, pp 423–440

Everett C, Madora K (2012) Quantity recognition among speakers of an Anumeric language. Cognit Sci 36(1):130

Greene JD (2016) Solving the trolley problem. In: Sytsma J, Buckwalter W (eds), pp 175–189

Haidt J (2001) The emotional dog and its rational tail: a social intuitionist approach to moral judgment. Psychol Rev 108:814–834

Jackson F (1982) Epiphenomenal qualia. Philos Quart 32:127–136

Jackson F (2011) 'On Gettier Holdouts'. Mind Lang 26(4):468–481.

Kahneman D (2011) Thinking, fast and slow. Farrar, Straus and Giroux, New York

Kauppinen A (2007) The rise and fall of experimental philosophy. Philos Explor 10(2):95–118. Reprinted In: Knobe J, Nichols S (eds), Experimental Philosophy, vol 2. Oxford: Oxford University Press 2014

Klampfer F (2017) The false promise of thought-experimentation in moral and political philosophy. In: Borstner B, Gartner S (eds), pp 328–348

Ludwig K (2018) Thought experiments and experimental philosophy. In: Stuart MT, Fehige Y, Brown JR (eds) The routledge companion to thought experiments. Routledge, pp 385–403

Machery E (2014) What is the significance of the demographic variation in semantic intuitions? In: Machery E, O'Neill E (eds) Current controversies in experimental philosophy. Routledge

Machery E (2017) Philosophy within its proper bounds. Oxford University Press, Oxford

Machery E, Mallon R, Nichols S, Stich S (2004) Semantics cross-cultural style. Cognition 92(3):B1–B12

Machery E, Stich S, Rose D, Chatterjee A, Karasawa K, Struchiner N, Sirker S, Usui N, Hashimoto T (2016) Gettier across cultures. Noûs 50(4)

Martin JW, Cushman W (2016) The adaptive logic of moral luck. In: Sytsma J, Buckwalter W (eds), pp 190–201

McPherson T (2015) The methodological irrelevance of reflective equilibrium. In Daly Ch (ed) The palgrave handbook of philosophical methods. Palgrave Macmillan, London

Mercier H, Sperber D (2017) The enigma of reason. Harvard University Press, Cambridge, Mass

Mortensen K, Nagel J (2016) Armchair-friendly experimental philosophy. In: Sytsma J, Buckwalter W (eds) A companion to experimental philosophy. Blackwell, Malden, MA, pp 53–70

Nahmias E, Morris S, Nadelhoffer T, Turner J (2005) Surveying freedom: folk intuitions about free will and moral responsibility. Philos Psychol 18(5):561–584

Nahmias E, Morris S, Nadelhoffer T, Turner J (2006) "Is incompatibilism intuitive? Res 73(1):28–53

Nersessian N (2007) Thought experiments as mental modeling. Croatian J Philos 7:125–161

Nichols S, Knobe J (2007) Moral responsibility and determinism: the cognitive science of folk intuitions. Nous 41(4):663–685

Nichols S, Stich S, Weinberg J (2003a) Metaskepticism: meditations in ethno-epistemology. In: Luper S (ed) The skeptics: contemporary debates. Ashgate, Burlington, pp 227–247

Nichols S, Stich S, Weinberg JM (2003b) Meta-skepticism: meditations in ethno-epistemology. In: Luper S (ed) The skeptics: contemporary essays. Ashgate, Aldershot

Nichols S, Stich S, Weinberg JM (2012) Meta-skepticism, meditations in ethno-epistemology. In Stephen stich collected papers, Volume 2: Knowledge, rationality, and morality, 1978–2010, first published In Luper S (ed) The Skeptics. Ashgate Publishing, Aldershot, England 2003, pp 227–247

Peacock KA (2018) Happiest thoughts great thought experiments of modern physics. In Stuart M, Fehige Y, Brown JR (eds) Routledge companion to thought experiments. Routledge, London, 211–241

Prinz J (2007) Can moral obligations be empirically discovered?. In: Wettstein H (ed) Midwest studies in philosophy 31

Rédei M (2003) Thinking about thought experiments in physics comment on Experiments and thought experiments in natural science. In: Galavotti MC (ed) Observation and experiment in the natural and social sciences. Springer Verlag, Switzerland, pp 237–241

Schnall S, Haidt J, Clore GL, Jordan AH (2008) Disgust as embodied moral judgment. Pers Soc Psychol Bull 34(8):1096–1109

Schütze CT (1986) The empirical base for linguistics, grammatical judgments and linguistic methodology. The University of Chicago Presss, Chicago

Sorensen RA (2014) Novice thought experiments. In: Booth AR, Rowbottom DP (eds) Intuitions. Oxford University Press, Oxford, pp 135–146

Sosa E (2007) "Experimental Philosophy and Philosophical Intuition." In: Knobe J, Nichols S (eds) Experimental philosophy. Oxford: Oxford University Press

Sosa E (2009) A defense of the use of intuitions in philosophy. In: Bishop M, Murphy D (eds) Stich and his critics. edited by, Blackwell Publishers, pp 101–112

Stenning K, Lambalgen Mv (2008) Human reasoning and cognitive science, a bradford book. The MIT Press, Cambridge, Massachusetts

Stich S, Tobia K (2018) Intuition and its critics. In: Stuart M, Fehige Y, Brown J (eds) Routledge handbook of thought experiments. Routledge, London, pp 369–383

Stich S, Mizumoto M, McCready E (2018) Introduction. In: Stich S, Mizumoto M, McCready E (eds) Epistemology for the rest of the world. Oxford University Press

Sytsma J, Livengood J (2011) New perspective concerning experiments on semantic intuitions. Australas J Philos 89(2):315–332

Swain S, Alexander J, Weinberg JM (2008) The instability of philosophical intuitions: running hot and cold on truetemp. Philos Phenomenol Res 76(1 January):138–155

Szabo GT (2010) Intuition, imagination, and philosophical methodology. Oxford University Press, Oxford

Weinberg JM (2007) How to change intuitions empirically without risking skepticism. Midwest Stud Philos 31 (1):318–343

Weinberg JM, Nichols Sh, Stich S (2001) Normativity and epistemic intuitions. Philos Top 29:429–460

Wierzbicka A (2018) I know a human universal. In: Stich S, Mizumoto M, McCready E (eds) Epistemology for the rest of the world. Oxford University Press, Oxford, pp 216–246

Williamson T (2007a) Philosophical knowledge and knowledge of counterfactuals. Grazer Philosophische Studien 74:89–123

Williamson T (2007b) The philosophy of philosophy. Blackwell Publishing, Oxford

Williamson T (2011) 'Philosophical expertise and the burden of proof'. Metaphilosophy 42(3):215–229

Williamson T (2016) Philosophical criticisms of experimental philosophy. [illegible]

Chapter 7
Conclusion

Let us wrap up, stressing the elements in the book that we find most important. We have seen that both in philosophy and science one encounters a lot of imaginative thought experiments. Both kinds, scientific and philosophical ones, often appeal to intuitions that typically result from such TEs. We have proposed that a TE is an investigative procedure "in the armchair", an imaginative enactment in thought, which normally involves the formulation of experimental design pointing to the determination of the goal(s),in particular the thesis/theory to be tested, and the construction of a (typically) counterfactual scenario to be considered. What follows is the presentation of the scenario thus constructed to the experimental subject (either the author of the scenario herself, or an interlocutor), the (typically imaginative) contemplation of the scenario and some piece of reasoning, and finally the decision ("intuition") concerning the thesis/theory to be tested. Often, the scenario is varied, and subject is invited to draw the conclusion from a series of answers to a series of varied scenarios; we used the term "intuitive induction" for this procedure. Normally, the conclusion is then compared and possibly contrasted to the dominant views in the field, from commonsensical to theoretical, scientific or philosophical ones. If all goes well a reflective equilibrium is reached…

There are also famous examples of TEs organized in a sequence, so as to yield a consistent narrative, and accompanied by relevant theorizing. In the philosophy of mind, we have a book-length TE, due to Condillac. In his *Treatise on Sensations* (from 1754) he asks the reader to identify oneself with a "statue" equipped with human-like neural apparatus (human-like robot, we would say nowadays), initially devoid of any contact with outside world. Its senses are awakened, one by one, and the reader is asked to imagine the reactions of the statue along its way. Also, I have mentioned groups of related micro-TEs integrated into a larger whole in Wittgenstein (most famously in his (1963) book),. The first thirty or so paragraphs of *Philosophical investigations* famously introduce various small "language games" (or proto-language games), commented and organized into an impressive whole, a true macro-TE. But, as we said in early chapters, the most dramatic examples come from political philosophy. The first is Plato's *Republic*. Socrates famously suggests at

N. Miscevic, *Thought Experiments*,
https://doi.org/10.1007/978-3-030-81082-5_7

[369a] to look for justice in states, and then only examine it also in the individual, and then proposes the following: "If, then," said I, "we should in *our argument* / logos / *observe the origin1 of a state*, we should see also the origin of justice and injustice in it." So, he and his interlocutors proceed to build the just state in their logos, by imagining a series of arrangements. Socrates, for instance, asks about community of children whether it is just, and his interlocutor, Adeimantos, imagines the arrangement, and volunteers the positive answer. Such a small imaginative experience might be treated as a mini-TE. Its contemporary counterpart is Rawls's *Theory of Justice*, where the veil of ignorace creates the scenario(s) needed, and an impressive political theory is built from questioning about principles of justice to be agreed to under the veil. So, big, integrative macro-TEs appear in some fundamental classical books of philosophy and are worth studying in more detail.

We have placed TEs on a wider map of other imaginative performances, stressing their relation to various interesting possible relatives: literary scenarios, meditative activities, political utopias and some others. We have proposed the name "imaginative enactments in thought", IETs for short, and stressed the need of comparative studying their traits.

How important are TEs? There is a consensus that they are very useful in science, in particular in introducing new ideas and helping to the birth of new theories: Galileo, Einstein and Heisenberg are the most cited cases, but Darwin's imaginative proposals in *The Origin of species* are also worth mentioning.[1]

In philosophy they seem to be indispensable, especially in the normative disciplines, like ethics and political philosophy. Real cases are simply not enough, neither to satisfy our moral curiosity, not to offer wide enough bases for generalizations, so TEs are needed as source of intuitions. We depend on imagined situations, and thereby on TEs to build, test and develop moral theories. Hence, in philosophy in general we simply cannot do without intuitions and TEs.

How are TEs to be explained? How can thinkers reach interesting and valid conclusion in the process? Here, we have been proposing mental modelling as the basic cognitive mechanism enabling experimenting in thought, distinguishing two kinds of it, the 3rd person model building and the 1st person simulation.

At this point one should note an interesting contrast concerning the generality of the approach: theoreticians that limit their research to, say, philosophical intuitions, tend to be divided equally between simulation-scenario and logic-inference approaches, whereas those who focus on say, linguistic or geometrical intuitions cannot but stress the specific role of simulation and imaginative scenarios. So, generality is on the side of the later: if we want to offer a more uniform account, we cannot get rid of simulation and scenarios; but once we introduce them, there is no reason not to rely on their importance wherever needed.

And we have proposed that the intuitions in which a TE normally results are typically products of various cognitive competences, from linguistic, through spatial and numerical to competences regulating our most general understanding of the

[1] Lennox (1991) and Schlaepfer and Weber (2018) for discussion of Lennox, and for more biological TEs.

world. Let me remind the reader briefly of the main ideas, and also place both proposals, the first one concerning mental modelling and the second, concerning specialized competences, on the map of options, present in the actual discussion.

I have stressed that modelling can make me understand in what way, by what causal means, some interesting and situation(-type) is possible. In chapter Four we used Einstein's Elevator TE, and here we might point to another example. Take Darwin's "just-so" stories concerning natural selection. (see references in fn.1) He asks:

> How do wolves develop to be slim and swift? Suppose that the fleetest prey, a deer for instance, had from any change in the country increased in numbers, or that other prey had decreased in numbers, during that season of the year when the wolf is hardest pressed for food. (Darwin 1859, pp. 90–91)

This will favor the swiftest and slimmest wolves, he suggests. Me, the reade, begin to understand that the swiftest and slimmest wolves would have the best chance of surviving, and so be preserved or selected. [...] A simple, movie-like mental model takes me from the scene in which the swiftest and slimmest wolf catches the dear, to the eating scene, and to the assumption that the wolf will be in good shape after its meal. And then females come; our wolf is full of energy and interest, and so on. The model makes me understand this first phase.

Now, my wolf has a good chance to leave offspring, and indeed more offspring than its heavy and slow competitor. And Darwin appeals to this fact:

> Now, if any slight innate change of habit or of structure benefited an individual wolf, it would have the best chance of surviving and of leaving offspring. Some of its young would probably inherit the same habits or structure, and by the repetition of this process, a new variety might be formed which would either supplant or coexist with the parent-form of wolf. (Darwin1859, pp. 90–91).

My mental model *depicts the causal structure, and thereby makes me consciously understand* how it works, both in short term (the catching and eating phase) and in long term (leaving more offspring that the competitors, who inherit at least some of their father's favorable traits). This is the case of *explanatory understanding* through model-building.[2]

We now turn to the crucial output of TEs, namely the intuitions. They are not final results of TEs, as we argued in Chapter Two: once an intuition is reached, there is further job of testing similar cases, aligning the intuitions obtained, reach an "intuitive conclusion", and then, if needed, confronting the intuition with other opinions, commonsensical or scientific, until a reflective equilibrium is reached. The nature of intuitions has exercised theoreticians of science, psychologists and philosophers for a very long time, at least since Descartes, if not since Aristotle. The two extreme views have been the following: the first one, which we named ordinarism, according to which they are just non–inferential beliefs, very much like

---

[2] On modelling and understanding see, for example Margaret Morrison 2009. She talks more about modelling in mathematical sense, but most of what she says is valid for models in general, mental models included.

other beliefs, and the second one, intuitional Platonism. A popular non-extreme variety is conceptualism, which sees intuitions as primarily concerned with concept (C. Peacocke). We have discussed the work of ordinarists (like Tim Williamson and Michael Devitt) and Platonists (like J.R. Brown): for a detailed discussion of conceptualism see Miscevic (2008), where I take Peacocke's views as my target and foil.

The proposal I came forward with Ie dubbed "the Moderate Voice-of-Competence proposal", using Devitt's ironic phrase Voice-of-Competence, and giving it a positive meaning. Authors like Austin and Chomsky have given specialized proposals along the same line, the first for general categorial intuitions, and the second for syntactic ones. I proposed a general picture: intuitions derive from specialized competences, arithmetical, spatial-geometrical, linguistic and various normative ones, e.g. moral and aesthetic competence. These are not predominantly conceptual, but involve sets of (unconscious) rules, or proto-theories that go way beyond merely specifying a concept. However, the "voice of competence" is often a discreet one, overlaid by influences from "central processor", and all sorts of external factors (general knowledge, interests, etc.); so it is often the task of the theorist to filter out the genuine product of competence. Reflection on intuitions (ordinary, specialist as well as philosophical) is holistic, and draws upon all sorts of beliefs and other attitudes.

This view of the genealogy of intuitions steers a middle course between the extremes of conceptualism, which acknowledges only conceptual capacities, and holistic ordinarism, which denies or minimizes the role of specialized competences in generating intuitions.

The second issue is, so to speak, definitory of the view taken here. Is intuition "special", as the tradition, and the contemporary mainstream would have it, or are "intuitions" mere judgment, completely ordinary …, as the opponents of the special status would claim. We should distinguish several dimensions of "being special" that often get confused in the heat of the polemic.

The one dimension concerns the contrast between dedicated domain-specific, and, domain-general abilities and processes. The first would be localistic, the second holistic. For instance, we often ask the following: are linguistic intuitions basically the product of dedicated, specialized linguistic competence, as Chomsky and his followers are claiming, or do they result crucially from the working of the central processor, that has access to the whole "web" of our beliefs, as Devitt would have it? Are moral intuitions basically engendered by "moral competence", or are they basically a mix, involving judgments of taste (disgusting vs. not disgusting-appetizing), judgments of social convention and all sort of other factors? Call the one side "special-dedicated" and the opposite side "non-special-holistic".

The other dimension is a within-domain contrast between intuitional and other judgments about the same target. Is the judgment in the imagined Gettier scenario different from a judgment in a real Gettier case? Is the recognition of linguistic (in-)correctness of an interlocutor's remark essentially different from "intuition" offered to an investigator, with the same remark proposed as a target in a corresponding question? Call the yes-side "special-imagined" and the no-side "non-special indifferent".

The third dimension is a within-domain contrast between extraordinary and ordinary objects of intuitional judgment. One might claim that the intuitional judgment is about, say, internal objects, for instance, internal representations, or special items like concepts, whereas ordinary judgments are about their purported usual objects. Call the one side "special-object" view, and the other "usual object view".

The distinctions are geared to present-day debates, and do not reflect all the contrasts important in history of the debate. For instance, on the traditional reading of Plato, the classical Platonic "intuition" (noesis) has a very specific, but extremely rich "object", namely the whole domain of ideas-Forms, so it is dedicated all right, and "domain-specific" in a different sense from the contemporary one (Forms are varied, metaphysical and moral, and they support a rich system of ideal mathematical objects—numbers, figures–akin to themselves).

My preferred view is the following: intuitions are basically products of "special-dedicated" competence(s). In a slogan, intuition is the Voice of Competence. However, a particular intuition(-state and -content) may contain "admixtures" of elements alien to competence. For example, a subject's linguistic intuitions may reflect her social prejudices or preferences, and her moral ones her selfish special pleading. As for the second contrast, I think that ordinary and intuitional judgments on the similar topic—say real Gettier vs. imagined Gettier ones—are not dramatically different. There might be some "special-imagined" quality to intuitions, namely that imagined situations and questioning tend to isolate practical consideration and focus subject's attention to the topic itself. The processing done by competence is like an off-line application of it, screening off practical interests, everyday routines.

Finally, my preference for extroverted, reference-focused approach leads me to reject the "special object" hypothesis and accept wholeheartedly the usual object view.

The resulting view is a pluralistic intuitionism-competentialism that I like to call Voice-of-Competence view, as I said above and in chapter Four; since it allows for admixtures to pure competence verdicts, I call it moderate Voice-of-Competence view. It is pluralistic in that it goes along with mainstream cognitive psychology in assuming a plurality of competences, each of them capable of offering crucial information for intuition. Next, we should suppose that there is a "public relations" capacity that presents the verdicts of the competence to the subject in the immediate, intuitional manner, very often with the degree of clarity, distinctness and attraction-compellingness that has been hailed by philosophical tradition as the mark of intuition. We have a choice to view it either as a particular "window" for "deep" competencies-capacities (linguistic, geometrical, "philosophical") or as capacity of first order capacity (= 2nd order capacity) to function in non-pragmatic, off line fashion.

Here I have proposed a view on intuitions elicited by thought experiment combining the following. First, intuitionism: there are intuitions-dispositions and judgments, which form a distinct group of phenomena, and there is the intuition-capacity, the capacity to use our imaginative and judgmental competencies in an off-line fashion. It is the voice of competence, most often discrete. Intuitional data are thus the minimal "products" of tentative production—linguistic, philosophical, moral or mathematical—by naïve thinker (or speaker-listener) and not their opinions about the data. The data involve no theory and very little proto-theory. Although there might be admixtures of guesswork in the conscious production of data, these are routinely weaned out by linguists. Finally, mere acceptance of the "voice of competence" does not land us in any objectionable Cartesianism: it is it is compatible with naturalism and with distrust of a priori philosophy.

Second, extroversion or referentialism: intuitions are concerned with their external objects, the domain of items and facts, rather than with concepts; here I agree with Tim Williamson (except that he does not want to call beliefs resulting from TEs "intuitions"). Concepts often play a role in the process, but they are not the object of intuitions, and their role is subordinate to the role played by the external referential domain.

Third, concerning TEs: they and the intuitions they produce are rather scenario-based than. inference-based. Imagining scenarios, typically particularized ones play the main cognitive and justificatory role, whereas inference typically plays a subordinate role. Fourth, concerning explanation, intuitions require an explanation of having and reliability, if possible a causal one. Fifth, concerning the justificatory status of intuitions, the stress is on a posteriori rather than on a priori.

Are all these competences realized by one and the same module or mechanism, or not? I follow the mainstream in taking seriously the reality of specific deficits: specific insults block specific abilities, so it is better to assume specialization of mechanisms. Linguistic competence is realized by some linguistic "box" in the head, spatial-geometrical by some "geometrical" box, moral intuitions are produced by a dedicated mechanism and so on.

With philosophy, a lot is unknown. The standard line in analytic tradition has been that philosophical intuitions derive from deep seated meanings or concepts through which thinkers conceptualize the basic features of reality (generating metaphysical intuitions), of cognitive values and norms (epistemological intuitions) and of morality. I said, and I apologize for repeating, that my own favorite line is summarized in the following paraphrase of J. L. Austin (where is term "words" is replaced with "concepts"): the stock of our deepest concepts embodies all the distinctions men have found worth drawing, and the connections they have found worth making, in the lifetimes of many generations: these surely are likely to be more numerous, more sound, since they have stood up to the long test of the survival of the fittest, and more subtle, at least in all ordinary and reasonably practical matters, than any that you or I are likely to think up on the spot (1980:. 182). This accumulated wisdom then allows the philosopher to anticipate the experience from the armchair. At the same time, the double fallibility of these intuitions accounts for the limits of philosophical autonomy: armchair research should be open to corrections from empirical science.

However, the decades of meaning-and-concept focused research have uncovered a degree of richness and flexibility of people's philosophical intuitions that don't go well with standard view(s) of concepts-definitions. On the one hand, philosophical intuitions seem to embrace more than just strict definitions of fundamental metaphysical, epistemological and moral categories (e.g. of IDENTITY, POSSIBILITY, KNOWLEDGE, OUGHT) and their narrow application. On the other hand, there is a margin for decision, change and disagreement, that doesn't fit the presumed rigidity of concepts. The standard conceptualist reaction is to introduce richer and more flexible mini-theories, often called conceptions, which fit the bill, and to ground philosophical intuitions in them, still abiding by the view of intuitions as results of "conceptual analysis". But calling theories "conceptions" doesn't make them conceptual; they are just deeply held systematic views about fundamental domains, like basic features of reality, plus cognitive and moral values and norms. So, why not talk about metaphysical, epistemological and moral competencies that work with fundamental theories about fundamental domains? To the extent that the core of these theories is very deeply held, possibly innate, indispensable for any cognitive and even practical project in the respective vast domain, the special status of philosophy would be preserved. To the extent that the core is ultimately revisable, and answerable to the facts about its object area, philosophy has to engage with empirical sciences and can and should be fitted into a naturalistic framework.

Also, our innate endowment might explain at least the very origin of the intuition-capacity and the initial stages of the formation of our intuition-states with their contents. For instance, it might consist of innate structures, some corresponding to concepts and some to inner, spatio-temporal "frames", responsible for an innate spatial-geometrical know-how to which Devitt is alluding in his considerations from analogy. It explains some of the objective validity of our intuitions. But nativism should be restricted to the origin of the system and to the relatively initial stages of processing. An intelligent nativist-adaptationist should allow for a wide margin of influence from individual empirical learning, which may overthrow even some deeply ingrained pre-conceptions to the contrary. And most importantly, intuition is doubly fallible. It can misrepresent the contents of our cognitive apparatus, and is thus internally fallible. But, the contents themselves–including their core, the innate assumptions–are also fallible, yielding the external fallibility of intuitions. Our innate geometry might be false, out possibly innate folk-physics certainly is. No deep or strong apriority is involved in their deliveries. In short, we can admit an important role of intuitions, and preserve some of their special status, intimated by their phenomenology, without falling into dangerous traps of classical Cartesianism. The appeal to innate endowment might help with the thorny issue of Benacerraf's Dilemma concerning ways we come in touch with abstract mathematical and modal properties: human beings have involved in the world which exemplifies mathematical and modal properties, and their evolution might have made them sensitive to such properties.

In the last two chapters we have talked about two important challenges of understanding of TEs. The first is understanding their history, and the second is the issue of how to improve our imperfect thought-experimental intuitions. We hope that attention to stages can help with both. As for history, it offers a connection between the inner structure of the TE and historical vicissitudes of traditions springing from it. This is typical for the understanding of the longer history of each given TE in the history of philosophy; for instance Evil demon or Social contract. These are long term histories, each lasting around three hundred years if we count only modern philosophy. If we could bring together short-term and long-term stages, this would make the historical process more intelligible. For instance, take the relatively recent variations of the two mentioned thought-experimental oldies. The Evil demon has been innovatively transformed into (or replaced with) the Brain-in-a vat and Social contract has generated the Original position proposal. What is the relation between such historical changes, and the usual micro-variations of a given TE? Once we have a worked out the answer we shall be able to integrate the meta-philosophy of TEs with information from history of philosophy and of theory of science.

Finally, there is a hope that attention to stages might teach us how to improve our thought-experimental reasoning, and thus minimize the kind of mistakes that experimental philosophy has been uncovering in the last two decades.

References

Darwin Ch (1859) On the origin of species. John Murray, London

Lennox JG (1991) Darwinian thought experiments: a function for just-so stories. In Horowitz T, Massey G (eds) Thought experiments in science and philosophy. Rowman and Littlefield, Savage

Miscevic N (2008) Competent voices: a theory of intuitions. http://oddelki.ff.uni-mb.si/filozofija

Schlaepfer G, Weber M (2018) Thought experiments in biology. In: Stuart M, Fehige Y, Brown JR (eds) Routledge companion to thought experiments. Routledge, London, pp 243–256

The manufacturer's authorised representative in the EU is Springer Nature Customer Service Centre GmbH, Europaplatz 3, 69115 Heidelberg, Germany. If you have any concerns regarding our products, please contact ProductSafety@springernature.com

Printed and bound by CPI Group (UK) Ltd, Croydon, CR0 4YY
30/09/2026
02225709-0004